TASC Subject Test

Mathematics

Student Practice Workbook

+ Two Realistic TASC Math Tests

Math Notion

www.MathNotion.com

TASC Subject Test – Mathematics

TASC Subject Test – Mathematics

TASC Subject Test Mathematics

Published in the United State of America By

The Math Notion

Web: WWW.MathNotion.com

Email: info@Mathnotion.com

Copyright © 2021 by the Math Notion. All rights reserved. No part of this publication may be reproduced, stored in a retrieval system, or transmitted in any form or by any means, electronic, mechanical, photocopying, recording, scanning, or otherwise, except as permitted under Section 107 or 108 of the 1976 United States Copyright Ac, without permission of the author.

All inquiries should be addressed to the Math Notion.

ISBN: 978-1-63620-044-6

TASC Subject Test – Mathematics

The Math Notion

Michael Smith has been a math instructor for over a decade now. He launched the Math Notion. Since 2006, we have devoted our time to both teaching and developing exceptional math learning materials. As a test prep company, we have worked with thousands of students. We have used the feedback of our students to develop a unique study program that can be used by students to drastically improve their math scores fast and effectively. We have more than a thousand Math learning books including:

– **GED Math Prep**

– **HiSET Math Prep**

– **TABE Math Prep**

– **Accuplacer Math Prep**

– **Common Core Math Prep**

–**many Math Education Workbooks, Study Guides, Practice and Exercise Books**

As an experienced Math test preparation company, we have helped many students raise their standardized test scores—and attend the colleges of their dreams: We tutor online and in person, we teach students in large groups, and we provide training materials and textbooks through our website and through Amazon.

You can contact us via email at:

info@Mathnotion.com

TASC Subject Test – Mathematics

Get the Targeted Practice You Need to Ace the TASC Math Test!

TASC Subject Test - Mathematics includes easy-to-follow instructions, helpful examples, and plenty of math practice problems to assist students to master each concept, brush up their problem-solving skills, and create confidence.

The TASC math practice book provides numerous opportunities to evaluate basic skills along with abundant remediation and intervention activities. It is a skill that permits you to quickly master intricate information and produce better leads in less time.

Students can boost their test-taking skills by taking the book's two practice TASC Math exams. All test questions answered and explained in detail.

Important Features of the TASC Math Book:

- A **complete review** of TASC math test topics,
- Over 2,500 practice problems covering all topics tested,
- The most important concepts you need to know,
- Clear and concise, easy-to-follow sections,
- Well designed for enhanced learning and interest,
- Hands-on experience with all question types
- **2 full-length practice tests** with detailed answer explanations
- Cost-Effective Pricing

Powerful math exercises to help you avoid traps and pacing yourself to beat the TASC test. Students will gain valuable experience and raise their confidence by taking math practice tests, learning about test structure, and gaining a deeper understanding of what is tested on the TASC Math. If ever there was a book to respond to the pressure to increase students' test scores, this is it.

TASC Subject Test – Mathematics

WWW.MathNotion.COM

… So Much More Online!

- ✓ FREE Math Lessons

- ✓ More Math Learning Books!

- ✓ Mathematics Worksheets

- ✓ Online Math Tutors

For a PDF Version of This Book

Please Visit WWW.MathNotion.com

TASC Subject Test – Mathematics

Contents

Chapter 1 : Integers and Number Theory 11
Rounding 12
Whole Number Addition and Subtraction 13
Whole Number Multiplication and Division 14
Rounding and Estimates 15
Adding and Subtracting Integers 16
Multiplying and Dividing Integers 17
Order of Operations 18
Ordering Integers and Numbers 19
Integers and Absolute Value 20
Factoring Numbers 21
Greatest Common Factor 22
Least Common Multiple 23
Answers of Worksheets 24

Chapter 2 : Fractions and Decimals 27
Simplifying Fractions 28
Adding and Subtracting Fractions 29
Multiplying and Dividing Fractions 30
Adding and Subtracting Mixed Numbers 31
Multiplying and Dividing Mixed Numbers 32
Adding and Subtracting Decimals 33
Multiplying and Dividing Decimals 34
Comparing Decimals 35
Rounding Decimals 36
Answers of Worksheets 37

Chapter 3 : Proportions, Ratios, and Percent 40
Simplifying Ratios 41
Proportional Ratios 42
Similarity and Ratios 43
Ratio and Rates Word Problems 44
Percentage Calculations 45
Percent Problems 46
Discount, Tax and Tip 47

WWW.MathNotion.Com

TASC Subject Test – Mathematics

Percent of Change 48
Simple Interest 49
Answers of Worksheets 50

Chapter 4 : Exponents and Radicals Expressions 53
Multiplication Property of Exponents 54
Zero and Negative Exponents 55
Division Property of Exponents 56
Powers of Products and Quotients 57
Negative Exponents and Negative Bases 58
Scientific Notation 59
Square Roots 60
Simplifying Radical Expressions 61
Answers of Worksheets 62

Chapter 5 : Algebraic Expressions 65
Simplifying Variable Expressions 66
Simplifying Polynomial Expressions 67
Translate Phrases into an Algebraic Statement 68
The Distributive Property 69
Evaluating One Variable Expressions 70
Evaluating Two Variables Expressions 71
Combining like Terms 72
Answers of Worksheets 73

Chapter 6 : Equations and Inequalities 75
One–Step Equations 76
Multi–Step Equations 77
Graphing Single-Variable Inequalities 78
One–Step Inequalities 79
Multi-Step Inequalities 80
Systems of Equations 81
Systems of Equations Word Problems 82
Answers of Worksheets 83

Chapter 7 : Linear Functions 87
Finding Slope 88
Graphing Lines Using Line Equation 89
Writing Linear Equations 90
Graphing Linear Inequalities 91

TASC Subject Test – Mathematics

Finding Midpoint 92
Finding Distance of Two Points 93
Answers of Worksheets 94

Chapter 8 : Polynomials 97
Writing Polynomials in Standard Form 98
Simplifying Polynomials 99
Adding and Subtracting Polynomials 100
Multiplying Monomials 101
Multiplying and Dividing Monomials 102
Multiplying a Polynomial and a Monomial 103
Multiplying Binomials 104
Factoring Trinomials 105
Operations with Polynomials 106
Answers of Worksheets 107

Chapter 9 : Functions Operations and Quadratic 111
Evaluating Function 112
Adding and Subtracting Functions 113
Multiplying and Dividing Functions 114
Composition of Functions 115
Quadratic Equation 116
Solving Quadratic Equations 117
Quadratic Formula and the Discriminant 118
Graphing Quadratic Functions 119
Answers of Worksheets 120

Chapter 10 : Geometry and Solid Figures 123
Angles 124
Pythagorean Relationship 125
Triangles 126
Polygons 127
Trapezoids 128
Circles 129
Cubes 130
Rectangular Prism 131
Cylinder 132
Pyramids and Cone 133
Answers of Worksheets 134

TASC Subject Test – Mathematics

Chapter 11 : Statistics and Probability 137
- Mean and Median 138
- Mode and Range 139
- Times Series 140
- Stem–and–Leaf Plot 141
- Pie Graph 142
- Probability Problems 143
- Answers of Worksheets 144

Chapter 12 : TASC Test Review 147
- TASC Practice Tests Answer Sheet 149
- TASC Practice Test 1 153
  - Section 1 - Calculator 153
  - Section 2 - No Calculator 163
- TASC Practice Test 2 167
  - Section 1 - Calculator 167
  - Section 2 - No Calculator 177

Chapter 13 : Answers and Explanations 181
- Answer Key 181
- Practice Test 1 183
  - Section 1- Calculator 183
  - Section 2 - No Calculator 190
- Practice Test 2 193
  - Section 1 - Calculator 193
  - Section 2- No Calculator 200

TASC Subject Test – Mathematics

Chapter 1:
Integers and Number Theory

Topics that you'll practice in this chapter:

- ✓ Rounding
- ✓ Whole Number Addition and Subtraction
- ✓ Whole Number Multiplication and Division
- ✓ Rounding and Estimates
- ✓ Adding and Subtracting Integers
- ✓ Multiplying and Dividing Integers
- ✓ Order of Operations
- ✓ Ordering Integers and Numbers
- ✓ Integers and Absolute Value
- ✓ Factoring Numbers
- ✓ Greatest Common Factor (GCF)
- ✓ Least Common Multiple (LCM)

"Wherever there is number, there is beauty." –Proclus

TASC Subject Test – Mathematics

Rounding

✎ **Round each number to the nearest ten.**

1) 42 = ___		5) 19 = ___		9) 48 = ___

2) 88 = ___		6) 25 = ___		10) 81 = ___

3) 24 = ___		7) 93 = ___		11) 58 = ___

4) 57 = ___		8) 71 = ___		12) 87 = ___

✎ **Round each number to the nearest hundred.**

13) 198 = ___		17) 321 = ___		21) 580 = ___

14) 387 = ___		18) 433 = ___		22) 868 = ___

15) 816 = ___		19) 579 = ___		23) 480 = ___

16) 101 = ___		20) 825 = ___		24) 287 = ___

✎ **Round each number to the nearest thousand.**

25) 1,382 = ___		29) 9,099 = ___		33) 52,866 = ___

26) 3,420 = ___		30) 22,980 = ___		34) 85,190 = ___

27) 4,254 = ___		31) 45,188 = ___		35) 70,990 = ___

28) 6,861 = ___		32) 16,808 = ___		36) 26,869 = ___

WWW.MathNotion.Com

TASC Subject Test – Mathematics

Whole Number Addition and Subtraction

✎ Find the sum or subtract.

1) 1,240
   + 658
   _____

2) 3,458
   − 544
   _____

3) 2,259
   − 752
   _____

4) 1,990
   + 324
   _____

5) 3,088
   + 229
   _____

6) 2,354
   +1,009
   _____

7) 2,855
   +4,455
   _____

8) 5,112
   +4,004
   _____

9) 4,822
   − 2,007
   _____

10) 8,380
    − 5,288
    _____

11) 3,227
    + 4,150
    _____

12) 7,702
    −4,331
    _____

✎ Find the missing number.

13) 720 + ____ = 1,360

14) 2,115 − ____ = 1,103

15) ____ + 3,105 = 6,200

16) 5,250 − 3,280 = ____

17) 8,020 + ____ = 8,990

18) 7,302 − 4,700 = ____

TASC Subject Test – Mathematics

Whole Number Multiplication and Division

✎ **Calculate each product.**

1) 42 × 13 = ____

2) 70 × 15 = ____

3) 40 × 14 = ____

4) 22 × 20 = ____

5) 110 × 11 = ____

6) 150 × 16 = ____

✎ **Find the missing quotient.**

7) 564 ÷ 6 = _____

8) 270 ÷ 3 = _____

9) 640 ÷ 8 = _____

10) 450 ÷ 9 = _____

11) 112 ÷ 7 = _____

12) 1,260 ÷ 9 = _____

13) 3,000 ÷ 10 = _____

14) 2,400 ÷ 8 = _____

15) 3,200 ÷ 40 = _____

16) 6,300 ÷ 90 = _____

✎ **Calculate each problem.**

17) 400 ÷ 5 = N, N = ___

18) 2,500 ÷ 10 = N, N = ___

19) N ÷ 3 = 150, N = ___

20) 42 × N = 252, N = ___

21) 660 ÷ N = 330, N = ___

22) N × 8 = 336, N = ___

WWW.MathNotion.Com

TASC Subject Test – Mathematics

Rounding and Estimates

✎ Estimate the sum by rounding each number to the nearest ten.

1) 13 + 22 = _____

2) 71 + 23 = _____

3) 61 + 58 = _____

4) 56 + 85 = _____

5) 368 + 249 = _____

6) 330 + 903 = _____

7) 471 + 293 = _____

8) 1,950 + 2,655 = _____

✎ Estimate the product by rounding each number to the nearest ten.

9) 32 × 71 = _____

10) 12 × 33 = _____

11) 31 × 83 = _____

12) 19 × 11 = _____

13) 42 × 76 = _____

14) 63 × 34 = _____

15) 19 × 31 = _____

16) 59 × 71 = _____

✎ Estimate the sum or product by rounding each number to the nearest ten.

17) 29 × 12 _____

18) 37 × 26 _____

19) 48 + 82 _____

20) 65 + 44 _____

21) 37 × 14 _____

22) 71 + 32 _____

WWW.MathNotion.Com

TASC Subject Test – Mathematics

Adding and Subtracting Integers

✍ **Find each sum.**

1) $14 + (-6) =$

2) $(-13) + (-20) =$

3) $5 + (-28) =$

4) $50 + (-12) =$

5) $(-7) + (-15) + 3 =$

6) $30 + (-14) + 8 =$

7) $40 + (-10) + (-14) + 17 =$

8) $(-15) + (-20) + 13 + 35 =$

9) $40 + (-20) + (38 - 29) =$

10) $28 + (-12) + (30 - 12) =$

✍ **Find each difference.**

11) $(-18) - (-7) =$

12) $25 - (-14) =$

13) $(-20) - 36 =$

14) $34 - (-19) =$

15) $51 - (30 - 21) =$

16) $17 - (5) - (-24) =$

17) $(35 + 20) - (-46) =$

18) $48 - 16 - (-8) =$

19) $62 - (28 + 17) - (-15) =$

20) $58 - (-23) - (-31) =$

21) $19 - (-8) - (-13) =$

22) $(19 - 24) - (-14) =$

23) $27 - 33 - (-21) =$

24) $58 - (32 + 24) - (-9) =$

25) $36 - (-30) + (-17) =$

26) $27 - (-42) + (-31) =$

TASC Subject Test – Mathematics

Multiplying and Dividing Integers

✍ **Find each product.**

1) $(-9) \times (-5) =$

2) $(-3) \times 9 =$

3) $8 \times (-12) =$

4) $(-7) \times (-20) =$

5) $(-3) \times (-5) \times 6 =$

6) $(14 - 3) \times (-8) =$

7) $12 \times (-9) \times (-3) =$

8) $(140 + 10) \times (-2) =$

9) $10 \times (-12 + 8) \times 3 =$

10) $(-8) \times (-5) \times (-10) =$

✍ **Find each quotient.**

11) $42 \div (-7) =$

12) $(-48) \div (-6) =$

13) $(-40) \div (-8) =$

14) $54 \div (-2) =$

15) $152 \div 19 =$

16) $(-144) \div (-12) =$

17) $180 \div (-10) =$

18) $(-312) \div (-12) =$

19) $221 \div (-13) =$

20) $(-126) \div (6) =$

21) $(-161) \div (-7) =$

22) $-266 \div (-14) =$

23) $(-120) \div (-4) =$

24) $270 \div (-18) =$

25) $(-208) \div (-8) =$

26) $(135) \div (-15) =$

WWW.MathNotion.Com

TASC Subject Test – Mathematics

Order of Operations

✎ **Evaluate each expression.**

1) $7 + (5 \times 4) =$

2) $14 - (3 \times 6) =$

3) $(19 \times 4) + 16 =$

4) $(16 - 7) - (8 \times 2) =$

5) $27 + (18 \div 3) =$

6) $(18 \times 8) \div 6 =$

7) $(32 \div 4) \times (-2) =$

8) $(9 \times 4) + (32 - 18) =$

9) $24 + (4 \times 3) + 7 =$

10) $(36 \times 3) \div (2 + 2) =$

11) $(-7) + (12 \times 3) + 11 =$

12) $(8 \times 5) - (24 \div 6) =$

13) $(7 \times 6 \div 3) - (12 + 9) =$

14) $(13 + 5 - 14) \times 3 - 2 =$

15) $(20 - 14 + 30) \times (64 \div 4) =$

16) $32 + (28 - (36 \div 9)) =$

17) $(7 + 6 - 4 - 7) + (15 \div 5) =$

18) $(85 - 20) + (20 - 18 + 7) =$

19) $(20 \times 2) + (14 \times 3) - 22 =$

20) $18 + 5 - (30 \times 3) + 20 =$

TASC Subject Test – Mathematics

Ordering Integers and Numbers

✎ **Order each set of integers from least to greatest.**

1) $8, -10, -5, -3, 4$ ___, ___, ___, ___, ___, ___

2) $-10, -18, 6, 14, 27$ ___, ___, ___, ___, ___, ___

3) $15, -8, -21, 21, -23$ ___, ___, ___, ___, ___, ___

4) $-14, -40, 23, -12, 47$ ___, ___, ___, ___, ___, ___

5) $59, -54, 32, -57, 36$ ___, ___, ___, ___, ___, ___

6) $68, 26, -19, 47, -34$ ___, ___, ___, ___, ___, ___

✎ **Order each set of integers from greatest to least.**

7) $18, 36, -16, -18, -10$ ___, ___, ___, ___, ___, ___

8) $27, 34, -12, -24, 94$ ___, ___, ___, ___, ___, ___

9) $50, -21, -13, 42, -2$ ___, ___, ___, ___, ___, ___

10) $37, 46, -20, -16, 86$ ___, ___, ___, ___, ___, ___

11) $-18, 88, -26, -59, 75$ ___, ___, ___, ___, ___, ___

12) $-65, -30, -25, 3, 14$ ___, ___, ___, ___, ___, ___

WWW.MathNotion.Com

TASC Subject Test – Mathematics

Integers and Absolute Value

✎ **Write absolute value of each number.**

1) $|-2| =$

2) $|-27| =$

3) $|-20| =$

4) $|14| =$

5) $|6| =$

6) $|-55| =$

7) $|16| =$

8) $|2| =$

9) $|54| =$

10) $|-4| =$

11) $|-11|$

12) $|88| =$

13) $|0| =$

14) $|79| =$

15) $|-32| =$

16) $|-17| =$

17) $|42| =$

18) $|-46| =$

19) $|1| =$

20) $|-40| =$

✎ **Evaluate the value.**

21) $|-5| - \frac{|-21|}{7} =$

22) $14 - |3 - 15| - |-4| =$

23) $\frac{|-32|}{4} \times |-4| =$

24) $\frac{|7 \times (-3)|}{7} \times \frac{|-19|}{3} =$

25) $|4 \times (-5)| + \frac{|-40|}{5} =$

26) $\frac{|-45|}{9} \times \frac{|-24|}{12} =$

27) $|-12 + 8| \times \frac{|-7 \times 7|}{7} =$

28) $\frac{|-11 \times 2|}{4} \times |-16| =$

WWW.MathNotion.Com

TASC Subject Test – Mathematics

Factoring Numbers

✏ **List all positive factors of each number.**

1) 9

2) 16

3) 24

4) 30

5) 26

6) 46

7) 20

8) 68

9) 28

10) 98

11) 14

12) 54

13) 55

14) 18

15) 63

16) 34

17) 50

18) 62

19) 95

20) 64

21) 70

22) 45

23) 22

24) 65

TASC Subject Test – Mathematics

Greatest Common Factor

✏ **Find the GCF for each number pair.**

1) 6, 2

2) 4, 5

3) 3, 12

4) 7, 3

5) 5, 10

6) 8, 48

7) 6, 18

8) 9, 15

9) 12, 18

10) 4, 36

11) 6, 10

12) 28, 52

13) 25, 10

14) 22, 24

15) 9, 54

16) 8, 54

17) 42, 14

18) 16, 40

19) 9, 2, 3

20) 5, 15, 10

21) 7, 9, 2

22) 16, 64

23) 30, 48

24) 36, 63

TASC Subject Test – Mathematics

Least Common Multiple

✎ Find the LCM for each number pair.

1) 6, 9

2) 15, 45

3) 16, 40

4) 12, 36

5) 18, 27

6) 14, 42

7) 6, 30

8) 8, 56

9) 7, 21

10) 8, 20

11) 15, 25

12) 7, 9

13) 4, 11

14) 8, 28

15) 28, 56

16) 40, 50

17) 12, 13

18) 22, 11

19) 36, 20

20) 15, 35

21) 18, 81

22) 30, 54

23) 18, 45

24) 75, 25

TASC Subject Test – Mathematics

Answers of Worksheets

Rounding

1) 40	10) 80	19) 600	28) 7,000
2) 90	11) 60	20) 800	29) 9,000
3) 20	12) 90	21) 600	30) 23,000
4) 60	13) 200	22) 900	31) 45,000
5) 20	14) 400	23) 500	32) 17,000
6) 30	15) 800	24) 300	33) 53,000
7) 90	16) 100	25) 1,000	34) 85,000
8) 70	17) 300	26) 3,000	35) 71,000
9) 50	18) 400	27) 4,000	36) 27,000

Whole Number Addition and Subtraction

1) 1,898	7) 7,310	13) 640
2) 2,914	8) 9,116	14) 1,012
3) 1,507	9) 2,815	15) 3,095
4) 2,314	10) 3,092	16) 1,970
5) 3,317	11) 7,377	17) 970
6) 3,363	12) 3,371	18) 2,602

Whole Number Multiplication and Division

1) 546	7) 94	13) 300	19) 450
2) 1,050	8) 90	14) 300	20) 6
3) 560	9) 80	15) 80	21) 2
4) 440	10) 50	16) 70	22) 42
5) 1,210	11) 16	17) 80	
6) 2,400	12) 140	18) 250	

Rounding and Estimates

1) 30	7) 760	13) 3,200	19) 130
2) 90	8) 4,610	14) 1,800	20) 110
3) 120	9) 2,100	15) 600	21) 400
4) 150	10) 300	16) 4,200	22) 100
5) 620	11) 2,400	17) 300	
6) 1,230	12) 200	18) 1,200	

WWW.MathNotion.Com

TASC Subject Test – Mathematics

Adding and Subtracting Integers

1) 8
2) −33
3) −23
4) 38
5) −19
6) 24
7) 33
8) 13
9) 29
10) 34
11) −11
12) 39
13) −56
14) 53
15) 42
16) 36
17) 101
18) 40
19) 32
20) 112
21) 40
22) 9
23) 15
24) 11
25) 49
26) 38

Multiplying and Dividing Integers

1) 45
2) −27
3) −96
4) 140
5) 90
6) −88
7) 324
8) −300
9) −120
10) −400
11) −6
12) 8
13) 5
14) −27
15) 8
16) 12
17) −18
18) 26
19) −17
20) −21
21) 23
22) 19
23) 30
24) −15
25) 26
26) −9

Order of Operations

1) 27
2) −4
3) 92
4) −7
5) 33
6) 24
7) −16
8) 50
9) 43
10) 27
11) 40
12) 36
13) −7
14) 10
15) 576
16) 56
17) 5
18) 74
19) 60
20) −47

Ordering Integers and Numbers

1) −10, −5, −3, 4, 8
2) −18, −10, 6, 14, 27
3) −23, −21, −8, 15, 21
4) −40, −14, −12, 23, 47
5) −57, −54, 32, 36, 59
6) −34, −19, 26, 47, 68
7) 36, 18, −10, −16, −18
8) 94, 34, 27, −12, −24
9) 50, 42, −2, −13, −21
10) 86, 46, 37, −16, −20
11) 88, 75, −18, −26, −59
12) 14, 3, −25, −30, −65

Integers and Absolute Value

1) 2
2) 27
3) 20
4) 14

WWW.MathNotion.Com

TASC Subject Test – Mathematics

5) 6	11) 11	17) 42	23) 32
6) 55	12) 88	18) 46	24) 19
7) 16	13) 0	19) 1	25) 28
8) 2	14) 79	20) 40	26) 10
9) 54	15) 32	21) 2	27) 28
10) 4	16) 17	22) -2	28) 88

Factoring Numbers

1) 1, 3, 9
2) 1, 2, 4, 8, 16
3) 1, 2, 3, 4, 6, 8, 12, 24
4) 1, 2, 3, 5, 6, 10, 15, 30
5) 1, 2, 13, 26
6) 1, 2, 23, 46
7) 1, 2, 4, 5, 10, 20
8) 1, 2, 4, 17, 34, 68
9) 1, 2, 4, 7, 14, 28
10) 1, 2, 7, 14, 49, 98
11) 1, 2, 7, 14
12) 1, 2, 3, 6, 9, 18, 27, 54
13) 1, 5, 11, 55
14) 1, 2, 3, 6, 9, 18
15) 1, 3, 7, 9, 21, 63
16) 1, 2, 17, 34
17) 1, 2, 5, 10, 25, 50
18) 1, 2, 31, 62
19) 1, 5, 19, 95
20) 1, 2, 4, 8, 16, 32, 64
21) 1, 2, 5, 7, 10, 14, 35, 70
22) 1, 3, 5, 9, 15, 45
23) 1, 2, 11, 22
24) 1, 5, 13, 65

Greatest Common Factor

1) 2	7) 6	13) 5	19) 1
2) 1	8) 3	14) 2	20) 5
3) 3	9) 6	15) 9	21) 1
4) 1	10) 4	16) 2	22) 16
5) 5	11) 2	17) 14	23) 6
6) 8	12) 4	18) 8	24) 9

Least Common Multiple

1) 18	7) 30	13) 44	19) 180
2) 45	8) 56	14) 56	20) 105
3) 80	9) 21	15) 56	21) 162
4) 36	10) 40	16) 200	22) 270
5) 54	11) 75	17) 156	23) 90
6) 42	12) 63	18) 22	24) 75

TASC Subject Test – Mathematics

Chapter 2 :
Fractions and Decimals

Topics that you'll practice in this chapter:

- ✓ Simplifying Fractions
- ✓ Adding and Subtracting Fractions
- ✓ Multiplying and Dividing Fractions
- ✓ Adding and Subtract Mixed Numbers
- ✓ Multiplying and Dividing Mixed Numbers
- ✓ Adding and Subtracting Decimals
- ✓ Multiplying and Dividing Decimals
- ✓ Comparing Decimals
- ✓ Rounding Decimals

"A Man is like a fraction whose numerator is what he is and whose denominator is what he thinks of himself. The larger the denominator, the smaller the fraction." –Tolstoy

TASC Subject Test – Mathematics

Simplifying Fractions

✍ **Simplify each fraction to its lowest terms.**

1) $\dfrac{5}{10} =$

2) $\dfrac{28}{35} =$

3) $\dfrac{27}{36} =$

4) $\dfrac{40}{80} =$

5) $\dfrac{14}{56} =$

6) $\dfrac{32}{48} =$

7) $\dfrac{52}{65} =$

8) $\dfrac{15}{60} =$

9) $\dfrac{80}{160} =$

10) $\dfrac{55}{77} =$

11) $\dfrac{28}{112} =$

12) $\dfrac{32}{64} =$

13) $\dfrac{63}{72} =$

14) $\dfrac{81}{90} =$

15) $\dfrac{35}{105} =$

16) $\dfrac{25}{70} =$

17) $\dfrac{80}{280} =$

18) $\dfrac{12}{81} =$

19) $\dfrac{36}{186} =$

20) $\dfrac{240}{540} =$

21) $\dfrac{70}{560} =$

✍ **Find the answer for each problem.**

22) Which of the following fractions equal to $\dfrac{3}{4}$? ____

    A. $\dfrac{60}{90}$ B. $\dfrac{43}{104}$ C. $\dfrac{48}{64}$ D. $\dfrac{150}{300}$

23) Which of the following fractions equal to $\dfrac{5}{8}$? ____

    A. $\dfrac{125}{200}$ B. $\dfrac{115}{200}$ C. $\dfrac{50}{100}$ D. $\dfrac{30}{90}$

24) Which of the following fractions equal to $\dfrac{3}{7}$? ____

    A. $\dfrac{58}{116}$ B. $\dfrac{54}{126}$ C. $\dfrac{270}{167}$ D. $\dfrac{42}{63}$

WWW.MathNotion.Com

TASC Subject Test – Mathematics

Adding and Subtracting Fractions

✍ **Find the sum.**

1) $\frac{5}{9} + \frac{4}{9} =$

2) $\frac{1}{2} + \frac{1}{7} =$

3) $\frac{3}{8} + \frac{1}{4} =$

4) $\frac{3}{5} + \frac{1}{2} =$

5) $\frac{1}{4} + \frac{3}{5} =$

6) $\frac{7}{8} + \frac{3}{8} =$

7) $\frac{1}{2} + \frac{7}{10} =$

8) $\frac{2}{5} + \frac{2}{3} =$

9) $\frac{5}{7} + \frac{2}{3} =$

10) $\frac{7}{12} + \frac{3}{4} =$

11) $\frac{5}{6} + \frac{2}{5} =$

12) $\frac{1}{12} + \frac{2}{3} =$

✍ **Find the difference.**

13) $\frac{1}{3} - \frac{1}{6} =$

14) $\frac{3}{4} - \frac{1}{8} =$

15) $\frac{1}{2} - \frac{1}{3} =$

16) $\frac{1}{4} - \frac{1}{5} =$

17) $\frac{5}{8} - \frac{2}{3} =$

18) $\frac{1}{4} - \frac{1}{7} =$

19) $\frac{5}{6} - \frac{1}{9} =$

20) $\frac{3}{4} - \frac{1}{6} =$

21) $\frac{7}{8} - \frac{1}{12} =$

22) $\frac{8}{15} - \frac{3}{5} =$

23) $\frac{3}{12} - \frac{1}{14} =$

24) $\frac{10}{13} - \frac{7}{26} =$

25) $\frac{6}{7} - \frac{3}{4} =$

26) $\frac{4}{5} - \frac{1}{8} =$

27) $\frac{4}{7} - \frac{2}{35} =$

28) $\frac{9}{16} - \frac{2}{8} =$

29) $\frac{8}{9} - \frac{7}{18} =$

30) $\frac{1}{2} - \frac{4}{9} =$

WWW.MathNotion.Com

TASC Subject Test – Mathematics

Multiplying and Dividing Fractions

✎ Find the value of each expression in lowest terms.

1) $\dfrac{1}{5} \times \dfrac{15}{5} =$

2) $\dfrac{9}{12} \times \dfrac{4}{9} =$

3) $\dfrac{1}{16} \times \dfrac{8}{10} =$

4) $\dfrac{1}{24} \times \dfrac{8}{10} =$

5) $\dfrac{1}{5} \times \dfrac{1}{4} =$

6) $\dfrac{7}{9} \times \dfrac{1}{7} =$

7) $\dfrac{6}{7} \times \dfrac{1}{3} =$

8) $\dfrac{2}{8} \times \dfrac{2}{8} =$

9) $\dfrac{5}{8} \times \dfrac{3}{5} =$

10) $\dfrac{4}{7} \times \dfrac{1}{8} =$

11) $\dfrac{7}{15} \times \dfrac{5}{7} =$

12) $\dfrac{3}{10} \times \dfrac{5}{9} =$

✎ Find the value of each expression in lowest terms.

13) $\dfrac{1}{4} \div \dfrac{1}{8} =$

14) $\dfrac{1}{10} \div \dfrac{1}{5} =$

15) $\dfrac{3}{4} \div \dfrac{1}{5} =$

16) $\dfrac{1}{3} \div \dfrac{5}{6} =$

17) $\dfrac{1}{7} \div \dfrac{8}{42} =$

18) $\dfrac{3}{4} \div \dfrac{1}{6} =$

19) $\dfrac{2}{7} \div \dfrac{7}{13} =$

20) $\dfrac{1}{24} \div \dfrac{3}{16} =$

21) $\dfrac{7}{12} \div \dfrac{5}{6} =$

22) $\dfrac{22}{18} \div \dfrac{11}{9} =$

23) $\dfrac{9}{35} \div \dfrac{3}{7} =$

24) $\dfrac{2}{7} \div \dfrac{8}{21} =$

25) $\dfrac{1}{9} \div \dfrac{2}{5} =$

26) $\dfrac{5}{12} \div \dfrac{3}{5} =$

27) $\dfrac{3}{20} \div \dfrac{1}{6} =$

28) $\dfrac{8}{20} \div \dfrac{3}{4} =$

29) $\dfrac{5}{6} \div \dfrac{2}{9} =$

30) $\dfrac{5}{11} \div \dfrac{3}{4} =$

WWW.MathNotion.Com

TASC Subject Test – Mathematics

Adding and Subtracting Mixed Numbers

✏️ **Find the sum.**

1) $3\frac{1}{3} + 2\frac{1}{6} =$

2) $4\frac{1}{2} + 3\frac{1}{2} =$

3) $3\frac{3}{8} + 1\frac{1}{8} =$

4) $2\frac{1}{4} + 2\frac{1}{3} =$

5) $3\frac{5}{6} + 2\frac{7}{12} =$

6) $5\frac{4}{15} + 3\frac{3}{5} =$

7) $2\frac{1}{3} + 4\frac{3}{7} =$

8) $3\frac{1}{2} + 4\frac{2}{5} =$

9) $5\frac{2}{5} + 6\frac{3}{7} =$

10) $8\frac{5}{16} + 6\frac{1}{12} =$

✏️ **Find the difference.**

11) $3\frac{1}{4} - 1\frac{3}{4} =$

12) $6\frac{3}{5} - 4\frac{2}{5} =$

13) $4\frac{1}{3} - 3\frac{1}{9} =$

14) $7\frac{1}{7} - 5\frac{1}{2} =$

15) $5\frac{1}{3} - 2\frac{1}{12} =$

16) $8\frac{1}{5} - 4\frac{1}{3} =$

17) $9\frac{1}{4} - 6\frac{1}{8} =$

18) $11\frac{7}{15} - 8\frac{3}{5} =$

19) $14\frac{5}{6} - 11\frac{3}{5} =$

20) $18\frac{2}{7} - 14\frac{1}{5} =$

21) $9\frac{1}{3} - 4\frac{1}{4} =$

22) $6\frac{1}{8} - 4\frac{1}{16} =$

23) $19\frac{3}{8} - 15\frac{1}{3} =$

24) $11\frac{1}{9} - 8\frac{1}{8} =$

25) $17\frac{1}{7} - 11\frac{1}{5} =$

26) $16\frac{2}{9} - 9\frac{5}{7} =$

WWW.MathNotion.Com

TASC Subject Test – Mathematics

Multiplying and Dividing Mixed Numbers

✎ **Find the product.**

1) $5\frac{1}{2} \times 2\frac{1}{4} =$

2) $5\frac{1}{3} \times 4\frac{1}{3} =$

3) $5\frac{3}{4} \times 6\frac{1}{4} =$

4) $3\frac{1}{3} \times 2\frac{3}{5} =$

5) $4\frac{8}{10} \times 1\frac{1}{24} =$

6) $6\frac{2}{7} \times 1\frac{1}{11} =$

7) $8\frac{2}{3} \times 3\frac{1}{2} =$

8) $3\frac{4}{7} \times 2\frac{1}{5} =$

9) $5\frac{2}{8} \times 4\frac{1}{6} =$

10) $7\frac{3}{3} \times 1\frac{3}{8} =$

✎ **Find the quotient.**

11) $2\frac{2}{5} \div 4\frac{1}{5} =$

12) $4\frac{1}{6} \div 3\frac{1}{3} =$

13) $6\frac{1}{3} \div 1\frac{1}{2} =$

14) $7\frac{1}{10} \div 2\frac{2}{5} =$

15) $3\frac{1}{3} \div 1\frac{1}{9} =$

16) $1\frac{1}{10} \div 4\frac{1}{2} =$

17) $1\frac{3}{16} \div 5\frac{1}{4} =$

18) $4\frac{1}{3} \div 4\frac{3}{4} =$

19) $9\frac{1}{3} \div 2\frac{1}{4} =$

20) $15\frac{1}{3} \div 5\frac{1}{2} =$

21) $4\frac{1}{6} \div 1\frac{1}{5} =$

22) $1\frac{1}{18} \div 1\frac{2}{9} =$

23) $4\frac{2}{7} \div 1\frac{3}{10} =$

24) $7\frac{1}{3} \div 2\frac{2}{11} =$

25) $8\frac{2}{5} \div 1\frac{1}{6} =$

26) $9\frac{1}{3} \div 2\frac{1}{7} =$

WWW.MathNotion.Com

TASC Subject Test – Mathematics

Adding and Subtracting Decimals

✎ Add and subtract decimals.

1) 35.19 − 24.28 = _____

4) 38.72 − 21.68 = _____

7) 86.09 − 35.14 = _____

2) 34.29 + 42.58 = _____

5) 57.39 + 26.54 = _____

8) 54.51 + 32.66 = _____

3) 61.20 + 33.75 = _____

6) 70.24 − 42.35 = _____

9) 114.21 − 88.69 = _____

✎ Find the missing number.

10) ___ + 2.8 = 5.4

11) 4.1 + ___ = 5.88

12) 6.45 + ___ = 8

13) 7.25 − ___ = 3.40

14) ___ − 2.35 = 4.25

15) ___ − 19.85 = 6.54

16) 22.15 + ___ = 28.95

17) ___ − 37.16 = 9.42

18) ___ + 24.50 = 34.19

19) 72.40 + ___ = 125.20

WWW.MathNotion.Com

TASC Subject Test – Mathematics

Multiplying and Dividing Decimals

✎ **Find the product.**

1) $0.5 \times 0.6 =$

2) $3.3 \times 0.4 =$

3) $1.28 \times 0.5 =$

4) $0.35 \times 0.6 =$

5) $1.85 \times 0.6 =$

6) $0.24 \times 0.5 =$

7) $5.25 \times 1.4 =$

8) $18.5 \times 4.6 =$

9) $15.4 \times 6.8 =$

10) $19.5 \times 2.6 =$

11) $32.2 \times 1.5 =$

12) $78.4 \times 4.5 =$

✎ **Find the quotient.**

13) $1.85 \div 10 =$

14) $74.6 \div 100 =$

15) $3.6 \div 3 =$

16) $9.6 \div 0.4 =$

17) $15.5 \div 0.5 =$

18) $32.8 \div 0.2 =$

19) $22.15 \div 1,000 =$

20) $53.55 \div 0.7 =$

21) $322.2 \div 0.2 =$

22) $50.67 \div 0.18 =$

23) $77.4 \div 0.8 =$

24) $27.93 \div 0.03 =$

TASC Subject Test – Mathematics

Comparing Decimals

✎ **Write the correct comparison symbol (>, < or =).**

1) 0.70 ☐ 0.070

2) 0.049 ☐ 0.49

3) 5.090 ☐ 5.09

4) 2.57 ☐ 2.05

5) 9.03 ☐ 0.930

6) 6.06 ☐ 6.6

7) 7.02 ☐ 7.020

8) 3.04 ☐ 3.2

9) 3.61 ☐ 3.245

10) 0.986 ☐ 0.0986

11) 17.24 ☐ 17.240

12) 0.759 ☐ 0.81

13) 9.040 ☐ 9.40

14) 5.73 ☐ 5.213

15) 9.44 ☐ 9.404

16) 7.17 ☐ 7.170

17) 4.85 ☐ 4.085

18) 9.041 ☐ 9.40

19) 3.033 ☐ 3.030

20) 4.97 ☐ 4.970

TASC Subject Test – Mathematics

Rounding Decimals

✎ **Round each decimal to the nearest whole number.**

1) 28.12 3) 16.22 5) 7.95

2) 6.9 4) 8.5 6) 52.7

✎ **Round each decimal to the nearest tenth.**

7) 31.761 9) 94.729 11) 13.219

8) 14.421 10) 77.89 12) 59.89

✎ **Round each decimal to the nearest hundredth.**

13) 8.428 15) 55.3786 17) 62.241

14) 23.812 16) 231.912 18) 19.447

✎ **Round each decimal to the nearest thousandth.**

19) 15.54324 21) 243.8652 23) 67.1983

20) 34.62586 22) 80.4529 24) 72.36788

TASC Subject Test – Mathematics

Answers of Worksheets

Simplifying Fractions

1) $\frac{1}{2}$
2) $\frac{4}{5}$
3) $\frac{3}{4}$
4) $\frac{1}{2}$
5) $\frac{1}{4}$
6) $\frac{2}{3}$
7) $\frac{4}{5}$
8) $\frac{1}{4}$
9) $\frac{1}{2}$
10) $\frac{5}{7}$
11) $\frac{1}{4}$
12) $\frac{1}{2}$
13) $\frac{7}{8}$
14) $\frac{9}{10}$
15) $\frac{1}{3}$
16) $\frac{5}{14}$
17) $\frac{2}{7}$
18) $\frac{4}{27}$
19) $\frac{6}{31}$
20) $\frac{4}{9}$
21) $\frac{1}{8}$
22) C
23) A
24) B

Adding and Subtracting Fractions

1) $\frac{9}{9} = 1$
2) $\frac{9}{14}$
3) $\frac{5}{8}$
4) $1\frac{1}{10}$
5) $\frac{17}{20}$
6) $1\frac{1}{4}$
7) $1\frac{1}{5}$
8) $1\frac{1}{15}$
9) $1\frac{8}{21}$
10) $1\frac{1}{3}$
11) $1\frac{7}{30}$
12) $\frac{3}{4}$
13) $\frac{1}{6}$
14) $\frac{5}{8}$
15) $\frac{1}{6}$
16) $\frac{1}{20}$
17) $-\frac{1}{24}$
18) $\frac{3}{28}$
19) $\frac{13}{18}$
20) $\frac{7}{12}$
21) $\frac{19}{24}$
22) $-\frac{1}{15}$
23) $\frac{5}{28}$
24) $\frac{1}{2}$
25) $\frac{3}{28}$
26) $\frac{27}{40}$
27) $\frac{18}{35}$
28) $\frac{5}{16}$
29) $\frac{1}{2}$
30) $\frac{1}{18}$

Multiplying and Dividing Fractions

1) $\frac{3}{5}$
2) $\frac{1}{3}$
3) $\frac{1}{20}$
4) $\frac{1}{30}$
5) $\frac{1}{20}$
6) $\frac{1}{9}$
7) $\frac{2}{7}$
8) $\frac{1}{16}$
9) $\frac{3}{8}$
10) $\frac{1}{14}$
11) $\frac{1}{3}$
12) $\frac{1}{6}$
13) 2
14) $\frac{1}{2}$

TASC Subject Test – Mathematics

15) $3\frac{3}{4}$

16) $\frac{2}{5}$

17) $\frac{3}{4}$

18) $4\frac{1}{2}$

19) $\frac{26}{49}$

20) $\frac{2}{9}$

21) $\frac{7}{10}$

22) 1

23) $\frac{3}{5}$

24) $\frac{3}{4}$

25) $\frac{5}{18}$

26) $\frac{25}{36}$

27) $\frac{9}{10}$

28) $\frac{8}{15}$

29) $3\frac{3}{4}$

30) $\frac{20}{33}$

Adding and Subtracting Mixed Numbers

1) $5\frac{1}{2}$

2) 8

3) $4\frac{1}{2}$

4) $4\frac{7}{12}$

5) $6\frac{5}{12}$

6) $8\frac{13}{15}$

7) $6\frac{16}{21}$

8) $7\frac{9}{10}$

9) $11\frac{29}{35}$

10) $14\frac{19}{48}$

11) $1\frac{1}{2}$

12) $2\frac{1}{5}$

13) $1\frac{2}{9}$

14) $1\frac{9}{14}$

15) $3\frac{1}{4}$

16) $3\frac{13}{15}$

17) $3\frac{1}{8}$

18) $2\frac{13}{15}$

19) $3\frac{7}{30}$

20) $4\frac{3}{35}$

21) $5\frac{1}{12}$

22) $2\frac{1}{16}$

23) $4\frac{1}{24}$

24) $2\frac{71}{72}$

25) $5\frac{33}{35}$

26) $6\frac{32}{63}$

Multiplying and Dividing Mixed Numbers

1) $12\frac{3}{8}$

2) $23\frac{1}{9}$

3) $35\frac{15}{16}$

4) $8\frac{2}{3}$

5) 5

6) $6\frac{6}{7}$

7) $30\frac{1}{3}$

8) $7\frac{6}{7}$

9) $21\frac{7}{8}$

10) 11

11) $\frac{4}{7}$

12) $1\frac{1}{4}$

13) $4\frac{2}{9}$

14) $2\frac{23}{24}$

15) 3

16) $\frac{11}{45}$

17) $\frac{19}{84}$

18) $\frac{52}{57}$

19) $4\frac{4}{27}$

20) $2\frac{26}{33}$

21) $3\frac{17}{36}$

22) $\frac{19}{22}$

23) $3\frac{27}{91}$

24) $3\frac{13}{36}$

25) $7\frac{1}{5}$

26) $4\frac{16}{45}$

Adding and Subtracting Decimals

1) 10.91

2) 76.87

3) 94.95

4) 17.04

TASC Subject Test – Mathematics

5) 83.93
6) 27.89
7) 50.95
8) 87.17
9) 25.52
10) 2.6
11) 1.78
12) 1.55
13) 3.85
14) 6.6
15) 26.39
16) 6.8
17) 46.58
18) 9.69
19) 52.8

Multiplying and Dividing Decimals

1) 0.3
2) 1.32
3) 0.64
4) 0.21
5) 1.11
6) 0.12
7) 7.35
8) 85.1
9) 104.72
10) 50.7
11) 48.3
12) 352.8
13) 0.185
14) 0.746
15) 1.2
16) 24
17) 31
18) 164
19) 0.02215
20) 76.5
21) 1,611
22) 281.5
23) 96.75
24) 931

Comparing Decimals

1) >
2) <
3) =
4) >
5) >
6) <
7) =
8) <
9) >
10) >
11) =
12) <
13) <
14) >
15) >
16) =
17) >
18) <
19) >
20) =

Rounding Decimals

1) 28
2) 7
3) 16
4) 9
5) 8
6) 53
7) 31.8
8) 14.4
9) 94.7
10) 77.9
11) 13.2
12) 59.9
13) 8.43
14) 23.81
15) 55.38
16) 231.91
17) 62.24
18) 19.45
19) 15.543
20) 34.626
21) 243.865
22) 80.453
23) 67.198
24) 72.368

WWW.MathNotion.Com

TASC Subject Test – Mathematics

Chapter 3 : Proportions, Ratios, and Percent

Topics that you'll practice in this chapter:

- ✓ Simplifying Ratios
- ✓ Proportional Ratios
- ✓ Similarity and Ratios
- ✓ Ratio and Rates Word Problems
- ✓ Percentage Calculations
- ✓ Percent Problems
- ✓ Discount, Tax and Tip
- ✓ Percent of Change
- ✓ Simple Interest

Without mathematics, there's nothing you can do. Everything around you is mathematics. Everything around you is numbers." – Shakuntala Devi

TASC Subject Test – Mathematics

Simplifying Ratios

✎ **Reduce each ratio.**

1) $15:20 = \_\_\_:\_\_\_$
2) $7:70 = \_\_\_:\_\_\_$
3) $16:28 = \_\_\_:\_\_\_$
4) $7:21 = \_\_\_:\_\_\_$
5) $4:40 = \_\_\_:\_\_\_$
6) $6:48 = \_\_\_:\_\_\_$
7) $16:64 = \_\_\_:\_\_\_$
8) $10:25 = \_\_\_:\_\_\_$
9) $8:48 = \_\_\_:\_\_\_$
10) $49:63 = \_\_\_:\_\_\_$
11) $18:27 = \_\_\_:\_\_\_$
12) $35:10 = \_\_\_:\_\_\_$
13) $90:9 = \_\_\_:\_\_\_$
14) $24:32 = \_\_\_:\_\_\_$
15) $7:56 = \_\_\_:\_\_\_$
16) $45:63 = \_\_\_:\_\_\_$
17) $56:72 = \_\_\_:\_\_\_$
18) $26:13 = \_\_\_:\_\_\_$
19) $15:45 = \_\_\_:\_\_\_$
20) $28:4 = \_\_\_:\_\_\_$
21) $24:48 = \_\_\_:\_\_\_$
22) $30:24 = \_\_\_:\_\_\_$
23) $70:140 = \_\_\_:\_\_\_$
24) $6:180 = \_\_\_:\_\_\_$

✎ **Write each ratio as a fraction in simplest form.**

25) $6:12 =$
26) $30:50 =$
27) $15:35 =$
28) $9:27 =$
29) $8:24 =$
30) $18:84 =$
31) $7:14 =$
32) $7:35 =$
33) $40:96 =$
34) $12:54 =$
35) $44:52 =$
36) $12:27 =$
37) $15:180 =$
38) $39:143 =$
39) $20:300 =$
40) $30:120 =$
41) $56:42 =$
42) $26:130 =$
43) $66:123 =$
44) $70:630 =$
45) $75:125 =$

TASC Subject Test – Mathematics

Proportional Ratios

✍ Fill in the blanks; Calculate each proportion.

1) $3:8 = \_\_ : 48$

2) $2:5 = 20: \_\_$

3) $1:9 = \_\_ : 81$

4) $6:7 = 12: \_\_$

5) $9:2 = 63: \_\_$

6) $8:7 = \_\_ : 49$

7) $20:3 = \_\_ : 15$

8) $1:3 = \_\_ : 75$

9) $7:6 = \_\_ : 60$

10) $8:5 = \_\_ : 45$

11) $3:10 = 60: \_\_$

12) $6:11 = 42: \_\_$

✍ State if each pair of ratios form a proportion.

13) $\frac{3}{20}$ and $\frac{9}{60}$

14) $\frac{1}{7}$ and $\frac{6}{42}$

15) $\frac{3}{7}$ and $\frac{24}{56}$

16) $\frac{4}{9}$ and $\frac{12}{18}$

17) $\frac{1}{9}$ and $\frac{12}{81}$

18) $\frac{7}{8}$ and $\frac{21}{28}$

19) $\frac{9}{13}$ and $\frac{27}{39}$

20) $\frac{1}{8}$ and $\frac{8}{64}$

21) $\frac{6}{19}$ and $\frac{30}{85}$

22) $\frac{5}{9}$ and $\frac{40}{81}$

23) $\frac{9}{14}$ and $\frac{108}{168}$

24) $\frac{15}{23}$ and $\frac{360}{552}$

✍ Calculate each proportion.

25) $\frac{20}{25} = \frac{32}{x}, x = \_\_\_$

26) $\frac{1}{8} = \frac{32}{x}, x = \_\_\_$

27) $\frac{15}{5} = \frac{21}{x}, x = \_\_\_$

28) $\frac{1}{7} = \frac{x}{294}, x = \_\_\_$

29) $\frac{7}{9} = \frac{x}{81}, x = \_\_\_$

30) $\frac{1}{5} = \frac{13}{x}, x = \_\_\_$

31) $\frac{9}{5} = \frac{36}{x}, x = \_\_\_$

32) $\frac{6}{13} = \frac{48}{x}, x = \_\_\_$

33) $\frac{5}{8} = \frac{x}{88}, x = \_\_\_$

34) $\frac{4}{15} = \frac{x}{240}, x = \_\_\_$

35) $\frac{9}{19} = \frac{x}{266}, x = \_\_\_$

36) $\frac{7}{15} = \frac{x}{270}, x = \_\_\_$

WWW.MathNotion.Com

TASC Subject Test – Mathematics

Similarity and Ratios

✎ **Each pair of figures is similar. Find the missing side.**

1) [triangles: 18, 12, ? and 6, 5, 4]

2) [triangles: 12, 9, 7.5 and 8, 6, ?]

3) [triangles: 44, 60 and 11, ?]

4) [trapezoids: 7, ? and 56, 104]

✎ **Calculate.**

5) Two rectangles are similar. The first is 24 feet wide and 120 feet long. The second is 30 feet wide. What is the length of the second rectangle? _____

6) Two rectangles are similar. One is 5 meters by 36 meters. The longer side of the second rectangle is 90 meters. What is the other side of the second rectangle? _____

7) A building casts a shadow 25 ft long. At the same time a girl 10 ft tall casts a shadow 5 ft long. How tall is the building? _____

8) The scale of a map of Texas is 4 inches: 32 miles. If you measure the distance from Dallas to Martin County as 38.4 inches, approximately how far is Martin County from Dallas? _____

TASC Subject Test – Mathematics

Ratio and Rates Word Problems

✎ Find the answer for each word problem.

1) Mason has 24 red cards and 36 green cards. What is the ratio of Mason 's red cards to his green cards? _____

2) In a party, 45 soft drinks are required for every 54 guests. If there are 378 guests, how many soft drinks is required? _____

3) In Mason's class, 42 of the students are tall and 24 are short. In Michael's class 84 students are tall and 48 students are short. Which class has a higher ratio of tall to short students? _____

4) The price of 5 apples at the Quick Market is $4.6. The price of 7 of the same apples at Walmart is $5.95. Which place is the better buy? _____

5) The bakers at a Bakery can make 90 bagels in 3 hours. How many bagels can they bake in 24 hours? What is that rate per hour? _____

6) You can buy 5 cans of green beans at a supermarket for $5.75. How much does it cost to buy 45 cans of green beans? _____

7) The ratio of boys to girls in a class is 4: 7. If there are 32 boys in the class, how many girls are in that class? _____

8) The ratio of red marbles to blue marbles in a bag is 3: 7. If there are 50 marbles in the bag, how many of the marbles are red? _____

WWW.MathNotion.Com

TASC Subject Test – Mathematics

Percentage Calculations

✎ **Calculate the given percent of each value.**

1) 3% of 60 = ____
2) 20% of 32 = ____
3) 4% of 72 = ____
4) 16% of 32 = ____
5) 25% of 124 = ____
6) 35% of 56 = ____

7) 15% of 20 = ____
8) 14% of 150 = ____
9) 80% of 50 = ____
10) 12% of 115 = ____
11) 72% of 250 = ____
12) 52% of 500 = ____

13) 70% of 400 = ____
14) 27% of 145 = ____
15) 90% of 64 = ____
16) 60% of 55 = ____
17) 22% of 210 = ____
18) 8% of 235 = ____

✎ **Calculate the percent of each given value.**

19) ____% of 25 = 5
20) ____% of 40 = 20
21) ____% of 25 = 2
22) ____% of 50 = 16
23) ____% of 250 = 5

24) ____% of 40 = 32
25) ____% of 125 = 20
26) ____% of 700 = 49
27) ____% of 350 = 49
28) ____% of 500 = 210

✎ **Calculate each percent problem.**

29) A Cinema has 250 seats. 60 seats were sold for the current movie. What percent of seats are empty? _____ %

30) There are 68 boys and 92 girls in a class. 75% of the students in the class take the bus to school. How many students do not take the bus to school? _____

WWW.MathNotion.Com

TASC Subject Test – Mathematics

Percent Problems

✎ **Calculate each problem.**

1) 9 is what percent of 45? ___%

2) 60 is what percent of 120? ___%

3) 10 is what percent of 200? ___%

4) 15 is what percent of 125? ___%

5) 10 is what percent of 400? ___%

6) 66 is what percent of 55? ___%

7) 40 is what percent of 160? ___%

8) 40 is what percent of 50? ___%

9) 120 is what percent of 800? ___%

10) 78 is what percent of 120? ___%

11) 36 is what percent of 144? ___%

12) 17 is what percent of 85? ___%

13) 90 is what percent of 900? ___%

14) 36 is what percent of 16? ___%

15) 63 is what percent of 14? ___%

16) 18 is what percent of 60? ___%

17) 126 is what percent of 200? ___%

18) 232 is what percent of 40? ___%

✎ **Calculate each percent word problem.**

19) There are 40 employees in a company. On a certain day, 25 were present. What percent showed up for work? ____%

20) A metal bar weighs 60 ounces. 25% of the bar is gold. How many ounces of gold are in the bar? _____

21) A crew is made up of 12 women; the rest are men. If 15% of the crew are women, how many people are in the crew? _____

22) There are 40 students in a class and 8 of them are girls. What percent are boys? ____%

23) The Royals softball team played 400 games and won 280 of them. What percent of the games did they lose? ____%

TASC Subject Test – Mathematics

Discount, Tax and Tip

✍ **Find the selling price of each item.**

1) Original price of a computer: $420
   Tax: 8% Selling price: $_____

2) Original price of a laptop: $280
   Tax: 4% Selling price: $_____

3) Original price of a sofa: $820
   Tax: 5% Selling price: $_____

4) Original price of a car: $15,800
   Tax: 3.6% Selling price: $_____

5) Original price of a Table: $250
   Tax: 9% Selling price: $_____

6) Original price of a house: $630,000
   Tax: 1.8% Selling price: $_____

7) Original price of a tablet: $450
   Discount: 30% Selling price: $____

8) Original price of a chair: $390
   Discount: 8% Selling price: $____

9) Original price of a book: $75
   Discount: 42% Selling price: $____

10) Original price of a cellphone: $820
    Discount: 23% Selling price: $___

11) Food bill: $45
    Tip: 15% Price: $_____

12) Food bill: $32
    Tipp: 20% Price: $_____

13) Food bill: $90
    Tip: 35% Price: $_____

14) Food bill: $42
    Tipp: 12% Price: $_____

✍ **Find the answer for each word problem.**

15) Nicolas hired a moving company. The company charged $500 for its services, and Nicolas gives the movers a 40% tip. How much does Nicolas tip the movers? $_____

16) Mason has lunch at a restaurant and the cost of his meal is $90. Mason wants to leave a 25% tip. What is Mason's total bill including tip? $_____

17) The sales tax in Texas is 19.80% and an item costs $350. How much is the tax? $_____

18) The price of a table at Best Buy is $680. If the sales tax is 5%, what is the final price of the table including tax? $_____

WWW.MathNotion.Com

TASC Subject Test – Mathematics

Percent of Change

🖎 **Find each percent of change.**

1) From 150 to 450. ___ %

2) From 50 ft to 250 ft. ___ %

3) From $60 to $360. ___ %

4) From 60 cm to 180 cm. ___ %

5) From 15 to 45. ___ %

6) From 80 to 16. ___ %

7) From 120 to 360. ___ %

8) From 900 to 450. ___ %

9) From 1,000 to 200. ___ %

10) From 144 to 36. ___ %

🖎 **Calculate each percent of change word problem.**

11) Bob got a raise, and his hourly wage increased from $42 to $63. What is the percent increase? ___ %

12) The price of a pair of shoes increases from $50 to $61. What is the percent increase? ___ %

13) At a coffee shop, the price of a cup of coffee increased from $4.80 to $5.76. What is the percent increase in the cost of the coffee? ___ %

14) 51 cm are cut from 85 cm board. What is the percent decrease in length? ___ %

15) In a class, the number of students has been increased from 54 to 81. What is the percent increase? ___ %

16) The price of gasoline rises from $24.40 to $30.50 in one month. By what percent did the gas price rise? ___ %

17) A shirt was originally priced at $38. It went on sale for $24.70. What was the percent that the shirt was discounted? ___ %

WWW.MathNotion.Com

TASC Subject Test – Mathematics

Simple Interest

✎ **Determine the simple interest for these loans.**

1) $480 at 11% for 3 years. $ _____

2) $4,200 at 7% for 4 years. $ _____

3) $2,500 at 20% for 3 years. $ _____

4) $6,800 at 3.9% for 4 months. $ _____

5) $800 at 6% for 7 months. $ _____

6) $36,000 at 4.2% for 6 years. $ _____

7) $6,500 at 7% for 4 years. $ _____

8) $850 at 9.5% for 2 years. $ _____

9) $1,200 at 5.8% for 9 months. $ _____

10) $3,000 at 4.5% for 7 years. $ _____

✎ **Calculate each simple interest word problem.**

11) A new car, valued at $22,000, depreciates at 8.5% per year. What is the value of the car one year after purchase? $_____

12) Sara puts $9,000 into an investment yielding 6% annual simple interest; she left the money in for three years. How much interest does Sara get at the end of those three years? $_____

13) A bank is offering 12% simple interest on a savings account. If you deposit $16,400, how much interest will you earn in two years? $_____

14) $720 interest is earned on a principal of $6,000 at a simple interest rate of 4% interest per year. For how many years was the principal invested? _____

15) In how many years will $2,200 yield an interest of $440 at 4% simple interest? _____

16) Jim invested $8,000 in a bond at a yearly rate of 4.5%. He earned $1,440 in interest. How long was the money invested? _____

WWW.MathNotion.Com

TASC Subject Test – Mathematics

Answers of Worksheets

Simplifying Ratios

1) 3 : 4
2) 1 : 10
3) 4 : 7
4) 1 : 3
5) 1 : 10
6) 1 : 8
7) 2 : 8
8) 2 : 5
9) 1 : 6
10) 7 : 9
11) 2 : 3
12) 7 : 2
13) 10 : 1
14) 3 : 4
15) 1 : 8
16) 5 : 7
17) 7 : 9
18) 2 : 1
19) 1 : 3
20) 7 : 1
21) 1 : 2
22) 5 : 4
23) 1 : 2
24) 1 : 30
25) $\frac{1}{2}$
26) $\frac{3}{5}$
27) $\frac{3}{7}$
28) $\frac{1}{3}$
29) $\frac{1}{3}$
30) $\frac{3}{14}$
31) $\frac{1}{2}$
32) $\frac{1}{5}$
33) $\frac{5}{12}$
34) $\frac{2}{9}$
35) $\frac{11}{13}$
36) $\frac{4}{9}$
37) $\frac{1}{12}$
38) $\frac{3}{11}$
39) $\frac{1}{15}$
40) $\frac{1}{4}$
41) $\frac{4}{3}$
42) $\frac{1}{5}$
43) $\frac{22}{41}$
44) $\frac{1}{9}$
45) $\frac{3}{5}$

Proportional Ratios

1) 18
2) 50
3) 9
4) 14
5) 14
6) 56
7) 100
8) 25
9) 70
10) 72
11) 200
12) 77
13) Yes
14) Yes
15) Yes
16) No
17) No
18) No
19) Yes
20) Yes
21) No
22) No
23) Yes
24) Yes
25) 40
26) 256
27) 7
28) 42
29) 63
30) 65
31) 20
32) 104
33) 55
34) 64
35) 126
36) 126

Similarity and ratios

1) 15
2) 5
3) 15
4) 13
5) 150 feet
6) 12.5 meters
7) 50 feet
8) 307.2 miles

Ratio and Rates Word Problems

1) 2 : 3
2) 315

WWW.MathNotion.Com

TASC Subject Test – Mathematics

3) The ratio for both classes is 7 to 4.
4) Walmart is a better buy.
5) 720, the rate is 30 per hour.

6) $51.75
7) 56
8) 15

Percentage Calculations

1) 1.8
2) 6.4
3) 2.88
4) 5.12
5) 31
6) 19.6
7) 3
8) 21
9) 40
10) 13.8

11) 180
12) 260
13) 280
14) 39.15
15) 57.6
16) 33
17) 46.2
18) 18.8
19) 20%
20) 50%

21) 8%
22) 32%
23) 2%
24) 80%
25) 16%
26) 7%
27) 14%
28) 42%
29) 76%
30) 40

Percent Problems

1) 20%
2) 50%
3) 5%
4) 12%
5) 2.5%
6) 120%
7) 25%
8) 80%

9) 15%
10) 65%
11) 25%
12) 20%
13) 10%
14) 225%
15) 450%
16) 30%

17) 63%
18) 580%
19) 62.5%
20) 15 ounces
21) 80
22) 80%
23) 30%

Discount, Tax and Tip

1) $453.60
2) $291.20
3) $861.00
4) $16,368.80
5) $272.50
6) $641,340

7) $315.00
8) $358.80
9) $43.50
10) $631.40
11) $51.75
12) $38.40

13) $121.50
14) $47.04
15) $200.00
16) $112.50
17) $69.30
18) $714.00

WWW.MathNotion.Com

TASC Subject Test – Mathematics

Percent of Change

1) 200%   7) 200%   13) 20%
2) 400%   8) 50%   14) 60%
3) 500%   9) 80%   15) 50%
4) 200%   10) 75%   16) 25%
5) 200%   11) 50%   17) 35%
6) 80%   12) 22%

Simple Interest

1) $158.40   7) $1,820.00   13) $3,936.00
2) $1,176.00   8) $161.50   14) 3 years
3) $1,500.00   9) $52.20   15) 5 years
4) $88.40   10) $945.00   16) 4 years
5) $28.00   11) $20,130.00
6) $9,072.00   12) $1,620.00

TASC Subject Test – Mathematics

Chapter 4 :
Exponents and Radicals Expressions

Topics that you'll practice in this chapter:

- ✓ Multiplication Property of Exponents
- ✓ Zero and Negative Exponents
- ✓ Division Property of Exponents
- ✓ Powers of Products and Quotients
- ✓ Negative Exponents and Negative Bases
- ✓ Scientific Notation
- ✓ Square Roots
- ✓ Simplifying Radical Expressions

Mathematics is no more computation than typing is literature.
– *John Allen Paulos*

TASC Subject Test – Mathematics

Multiplication Property of Exponents

✎ Simplify and write the answer in exponential form.

1) $4 \times 4^5 =$

2) $8^4 \times 8 =$

3) $7^3 \times 7^3 =$

4) $9^2 \times 9^2 =$

5) $2^2 \times 2^4 \times 2 =$

6) $5 \times 5^3 \times 5^3 =$

7) $4^3 \times 4^2 \times 4 \times 4 =$

8) $5x \times x =$

9) $x^3 \times x^3 =$

10) $x^7 \times x^2 =$

11) $x^4 \times x^3 \times x^2 =$

12) $10x \times 3x =$

13) $4x^3 \times 4x^3 =$

14) $7x^3 \times x =$

15) $3x^2 \times 4x^2 \times x^2 =$

16) $5x^4 \times x^4 =$

17) $2x^8 \times 2x =$

18) $6x \times x^5 =$

19) $4x^2 \times 6x^6 =$

20) $5yx^3 \times 4x =$

21) $7x^3 \times y^5 x^7 =$

22) $y^2 x^3 \times y^5 x^4 =$

23) $3x^5 \times 4x^3 y^4 =$

24) $4x^4 \times 9x^2 y^5 =$

25) $5x^3 y^4 \times 6x^8 y^2 =$

26) $8x^3 y^6 \times 4xy^3 =$

27) $2xy^5 \times 6x^3 y^3 =$

28) $4x^5 y^2 \times 4x^2 y^8 =$

29) $7x \times 3y^8 x^2 \times y^5 =$

30) $x^3 \times 2y^3 x^4 \times 2y =$

31) $3yx^4 \times 3y^4 x \times 3xy^3 =$

32) $6y^3 \times 2y^2 x^4 \times 10yx^5 =$

WWW.MathNotion.Com

TASC Subject Test – Mathematics

Zero and Negative Exponents

✎ **Evaluate the following expressions.**

1) $1^{-5} =$

2) $4^{-1} =$

3) $0^{10} =$

4) $1^{15} =$

5) $5^{-2} =$

6) $3^{-3} =$

7) $9^{-1} =$

8) $10^{-2} =$

9) $12^{-2} =$

10) $2^{-5} =$

11) $3^{-4} =$

12) $2^{-4} =$

13) $6^{-3} =$

14) $10^{-3} =$

15) $30^{-1} =$

16) $15^{-2} =$

17) $4^{-3} =$

18) $2^{-7} =$

19) $5^{-3} =$

20) $4^{-4} =$

21) $3^{-5} =$

22) $10^{-4} =$

23) $2^{-10} =$

24) $8^{-3} =$

25) $20^{-2} =$

26) $14^{-2} =$

27) $9^{-3} =$

28) $100^{-2} =$

29) $5^{-4} =$

30) $4^{-6} =$

31) $\left(\frac{1}{4}\right)^{-3} =$

32) $\left(\frac{1}{6}\right)^{-2} =$

33) $\left(\frac{1}{7}\right)^{-2} =$

34) $\left(\frac{2}{3}\right)^{-3} =$

35) $\left(\frac{1}{13}\right)^{-2} =$

36) $\left(\frac{7}{12}\right)^{-2} =$

37) $\left(\frac{1}{6}\right)^{-3} =$

38) $\left(\frac{1}{300}\right)^{-2} =$

39) $\left(\frac{2}{9}\right)^{-2} =$

40) $\left(\frac{7}{5}\right)^{-1} =$

41) $\left(\frac{13}{23}\right)^{0} =$

42) $\left(\frac{1}{4}\right)^{-5} =$

WWW.MathNotion.Com

TASC Subject Test – Mathematics

Division Property of Exponents

✎ Simplify.

1) $\dfrac{5^6}{5^7} =$

2) $\dfrac{8^8}{8^6} =$

3) $\dfrac{4^5}{4} =$

4) $\dfrac{3}{3^5} =$

5) $\dfrac{x}{x^6} =$

6) $\dfrac{3 \times 3^2}{3^2 \times 3^5} =$

7) $\dfrac{9^4}{9^2} =$

8) $\dfrac{10 \times 10^9}{10^2 \times 10^7} =$

9) $\dfrac{7^5 \times 7^7}{7^4 \times 7^8} =$

10) $\dfrac{15x}{30x^6} =$

11) $\dfrac{3x^9}{4x^4} =$

12) $\dfrac{15x^8}{10x^9} =$

13) $\dfrac{42x^5}{6y^9} =$

14) $\dfrac{36y^8}{4x^4y^5} =$

15) $\dfrac{2x^7}{9x} =$

16) $\dfrac{49x^8y^6}{7x^9} =$

17) $\dfrac{48x^2}{24x^6y^{12}} =$

18) $\dfrac{30yx^5}{6yx^7} =$

19) $\dfrac{19x^7y}{38x^{12}y^4} =$

20) $\dfrac{9x^8}{63x^8} =$

21) $\dfrac{9x^{-9}}{4x^{-3}} =$

WWW.MathNotion.Com

Powers of Products and Quotients

✎ **Simplify.**

1) $(4^3)^2 =$

2) $(2^3)^4 =$

3) $(2 \times 2^3)^2 =$

4) $(5 \times 5^5)^6 =$

5) $(19^4 \times 19^2)^3 =$

6) $(2^3 \times 2^4)^4 =$

7) $(5 \times 5^2)^2 =$

8) $(4^4)^4 =$

9) $(8x^5)^2 =$

10) $(3x^2y^4)^4 =$

11) $(7x^5y^2)^2 =$

12) $(5x^4y^4)^3 =$

13) $(2x^3y^3)^5 =$

14) $(10x^3y^4)^3 =$

15) $(13y^3y)^2 =$

16) $(5x^6x^4)^2 =$

17) $(6x^7y^6)^3 =$

18) $(12x^5x^7)^2 =$

19) $(2x^4 \times 2x)^4 =$

20) $(2x^4y^3)^5 =$

21) $(15x^7y^2)^2 =$

22) $(8x^3y^5)^3 =$

23) $(3x \times 2y^2)^4 =$

24) $\left(\dfrac{4x}{x^5}\right)^2 =$

25) $\left(\dfrac{x^4y^5}{x^3y^5}\right)^9 =$

26) $\left(\dfrac{36xy}{6x^5}\right)^3 =$

27) $\left(\dfrac{x^7}{x^8y^2}\right)^6 =$

28) $\left(\dfrac{xy^4}{x^3y^6}\right)^{-3} =$

29) $\left(\dfrac{5xy^8}{x^3}\right)^2 =$

30) $\left(\dfrac{xy^6}{2xy^3}\right)^{-4} =$

TASC Subject Test – Mathematics

Negative Exponents and Negative Bases

✎ **Simplify.**

1) $-9^{-1} =$

2) $-9^{-2} =$

3) $-2^{-5} =$

4) $-x^{-7} =$

5) $11x^{-1} =$

6) $-8x^{-3} =$

7) $-12x^{-5} =$

8) $-9x^{-8}y^{-6} =$

9) $32x^{-5}y^{-1} =$

10) $10a^{-9}b^{-3} =$

11) $-17x^4y^{-6} =$

12) $-\dfrac{25}{x^{-5}} =$

13) $-\dfrac{13x}{a^{-7}} =$

14) $\left(-\dfrac{1}{3}\right)^{-4} =$

15) $\left(-\dfrac{3}{4}\right)^{-2} =$

16) $-\dfrac{14}{a^{-6}b^{-3}} =$

17) $-\dfrac{7x}{x^{-8}} =$

18) $-\dfrac{a^{-9}}{b^{-5}} =$

19) $-\dfrac{11}{x^{-5}} =$

20) $\dfrac{8b}{-16c^{-6}} =$

21) $\dfrac{12ab}{a^{-4}b^{-3}} =$

22) $-\dfrac{8n^{-4}}{32p^{-7}} =$

23) $\dfrac{16ab^{-6}}{-6c^{-5}} =$

24) $\left(\dfrac{10a}{5c}\right)^{-4} =$

25) $\left(-\dfrac{12x}{4yz}\right)^{-3} =$

26) $\dfrac{8ab^{-7}}{-5c^{-3}} =$

27) $\left(-\dfrac{x^4}{x^5}\right)^{-5} =$

28) $\left(-\dfrac{x^{-2}}{7x^3}\right)^{-2} =$

29) $\left(-\dfrac{x^{-4}}{x^2}\right)^{-6} =$

WWW.MathNotion.Com

TASC Subject Test – Mathematics

Scientific Notation

✎ Write each number in scientific notation.

1) 0.223 =

2) 0.09 =

3) 4.5 =

4) 900 =

5) 2,000 =

6) 0.006 =

7) 33 =

8) 9,400 =

9) 1,470 =

10) 52,000 =

11) 8,000,000 =

12) 0.00009 =

13) 2,158,000 =

14) 0.0039 =

15) 0.000075 =

16) 4,300,000 =

17) 130,000 =

18) 4,000,000,000 =

19) 0.00009 =

20) 0.0039 =

✎ Write each number in standard notation.

21) 4×10^{-1} =

22) 1.2×10^{-3} =

23) 2.7×10^{5} =

24) 6×10^{-4} =

25) 3.6×10^{-3} =

26) 5.5×10^{5} =

27) 3.2×10^{4} =

28) 3.88×10^{6} =

29) 7×10^{-6} =

30) 4.2×10^{-7} =

TASC Subject Test – Mathematics

Square Roots

✎ **Find the value each square root.**

1) $\sqrt{16}$ = ___

2) $\sqrt{25}$ = ___

3) $\sqrt{1}$ = ___

4) $\sqrt{64}$ = ___

5) $\sqrt{0}$ = ___

6) $\sqrt{196}$ = ___

7) $\sqrt{4}$ = ___

8) $\sqrt{256}$ = ___

9) $\sqrt{36}$ = ___

10) $\sqrt{289}$ = ___

11) $\sqrt{169}$ = ___

12) $\sqrt{144}$ = ___

13) $\sqrt{100}$ = ___

14) $\sqrt{1,600}$ = ___

15) $\sqrt{2,500}$ = ___

16) $\sqrt{324}$ = ___

17) $\sqrt{529}$ = ___

18) $\sqrt{20}$ = ___

19) $\sqrt{625}$ = ___

20) $\sqrt{18}$ = ___

21) $\sqrt{50}$ = ___

22) $\sqrt{1,024}$ = ___

23) $\sqrt{160}$ = ___

24) $\sqrt{32}$ = ___

✎ **Evaluate.**

25) $\sqrt{4} \times \sqrt{25}$ = ___

26) $\sqrt{36} \times \sqrt{49}$ = ___

27) $\sqrt{6} \times \sqrt{6}$ = ___

28) $\sqrt{13} \times \sqrt{13}$ = ___

29) $2\sqrt{5} \times 3\sqrt{5}$ = ___

30) $\sqrt{12} \times \sqrt{3}$ = ___

31) $\sqrt{13} + \sqrt{13}$ = ___

32) $\sqrt{10} + 2\sqrt{10}$ = ___

33) $12\sqrt{7} - 10\sqrt{7}$ = ___

34) $4\sqrt{10} \times 2\sqrt{10}$ = ___

35) $5\sqrt{3} \times 8\sqrt{3}$ = ___

36) $6\sqrt{3} - \sqrt{12}$ = ___

TASC Subject Test – Mathematics

Simplifying Radical Expressions

✎ **Simplify.**

1) $\sqrt{13x^2} =$

2) $\sqrt{75x^2} =$

3) $\sqrt[3]{27a} =$

4) $\sqrt{64x^5} =$

5) $\sqrt{216a} =$

6) $\sqrt[3]{63w^3} =$

7) $\sqrt{192x} =$

8) $\sqrt{125v} =$

9) $\sqrt[3]{128x^2} =$

10) $\sqrt{100x^9} =$

11) $\sqrt{16x^4} =$

12) $\sqrt[3]{500a^5} =$

13) $\sqrt{242} =$

14) $\sqrt{392p^3} =$

15) $\sqrt{8m^6} =$

16) $\sqrt{198x^3y^3} =$

17) $\sqrt{121x^5y^5} =$

18) $\sqrt{16a^6b^3} =$

19) $\sqrt{90x^5y^7} =$

20) $\sqrt[3]{64y^2x^6} =$

21) $10\sqrt{16x^4} =$

22) $6\sqrt{81x^2} =$

23) $\sqrt[3]{56x^2y^6} =$

24) $\sqrt[3]{1,000x^5y^7} =$

25) $8\sqrt{50a} =$

26) $\sqrt[4]{625x^8y} =$

27) $\sqrt{24x^4y^5r^3} =$

28) $5\sqrt{36x^4y^5z^8} =$

29) $3\sqrt[3]{343x^9y^7} =$

30) $5\sqrt{81a^5b^2c^9} =$

31) $\sqrt[4]{625x^8y^{16}} =$

WWW.MathNotion.Com

TASC Subject Test – Mathematics

Answers of Worksheets

Multiplication Property of Exponents

1) 4^6
2) 8^5
3) 7^6
4) 9^4
5) 2^7
6) 5^7
7) 4^7
8) $5x^2$
9) x^6
10) x^9
11) x^9
12) $30x^2$
13) $16x^6$
14) $7x^4$
15) $12x^6$
16) $5x^8$
17) $4x^9$
18) $6x^6$
19) $24x^8$
20) $20x^4y$
21) $7x^{10}y^5$
22) x^7y^7
23) $12x^8y^4$
24) $36x^6y^5$
25) $30x^{11}y^6$
26) $32x^4y^9$
27) $12x^4y^8$
28) $16x^7y^{10}$
29) $21x^3y^{13}$
30) $4x^7y^4$
31) $27x^6y^8$
32) $120x^9y^6$

Zero and Negative Exponents

1) 1
2) $\frac{1}{4}$
3) 0
4) 1
5) $\frac{1}{25}$
6) $\frac{1}{27}$
7) $\frac{1}{9}$
8) $\frac{1}{100}$
9) $\frac{1}{144}$
10) $\frac{1}{32}$
11) $\frac{1}{81}$
12) $\frac{1}{16}$
13) $\frac{1}{216}$
14) $\frac{1}{1,000}$
15) $\frac{1}{30}$
16) $\frac{1}{225}$
17) $\frac{1}{64}$
18) $\frac{1}{128}$
19) $\frac{1}{125}$
20) $\frac{1}{256}$
21) $\frac{1}{243}$
22) $\frac{1}{10,000}$
23) $\frac{1}{1,024}$
24) $\frac{1}{512}$
25) $\frac{1}{400}$
26) $\frac{1}{196}$
27) $\frac{1}{729}$
28) $\frac{1}{10,000}$
29) $\frac{1}{625}$
30) $\frac{1}{4,096}$
31) 64
32) 36
33) 49
34) $\frac{27}{8}$
35) 169
36) $\frac{144}{49}$
37) 216
38) $90,000$
39) $\frac{81}{4}$
40) $\frac{5}{7}$
41) 1
42) $1,024$

Division Property of Exponents

1) $\frac{1}{5}$
2) 8^2
3) 4^4
4) $\frac{1}{3^4}$
5) $\frac{1}{x^5}$
6) $\frac{1}{3^4}$
7) 9^2
8) 10
9) 1
10) $\frac{1}{2x^5}$
11) $\frac{3x^5}{4}$
12) $\frac{3}{2x}$
13) $\frac{7x^5}{y^9}$
14) $\frac{9y^3}{x^4}$

WWW.MathNotion.Com

TASC Subject Test – Mathematics

15) $\frac{2x^6}{9}$ 17) $\frac{2}{x^4 y^{12}}$ 19) $\frac{1}{2x^5 y^3}$ 21) $\frac{9}{4x^6}$

16) $\frac{7y^6}{x}$ 18) $\frac{5}{x^2}$ 20) $\frac{1}{7}$

Powers of Products and Quotients

1) 4^6 12) $125x^{12}y^{12}$ 23) $1{,}296x^4 y^8$

2) 2^{12} 13) $32x^{15}y^{15}$ 24) $\frac{16}{x^8}$

3) 2^8 14) $1{,}000x^9 y^{12}$ 25) x^9

4) 5^{36} 15) $169y^8$ 26) $\frac{216y^3}{x^{12}}$

5) 19^{18} 16) $25x^{20}$ 27) $\frac{1}{x^6 y^{12}}$

6) 2^{28} 17) $216x^{21}y^{18}$ 28) $x^6 y^6$

7) 5^6 18) $144x^{24}$ 29) $\frac{25y^{16}}{x^4}$

8) 4^{16} 19) $256x^{20}$

9) $64x^{10}$ 20) $32x^{20}y^{15}$ 30) $\frac{16}{y^{12}}$

10) $81x^8 y^{16}$ 21) $225x^{14}y^4$

11) $49x^{10}y^4$ 22) $512x^9 y^{15}$

Negative Exponents and Negative Bases

1) $-\frac{1}{9}$ 8) $-\frac{9}{x^8 y^6}$ 16) $-14a^6 b^3$ 24) $\frac{c^4}{16a^4}$

2) $-\frac{1}{81}$ 9) $\frac{32}{x^5 y}$ 17) $-7x^9$ 25) $\frac{y^3 z^3}{27x^3}$

3) $-\frac{1}{32}$ 10) $\frac{10}{a^9 b^3}$ 18) $-\frac{b^5}{a^9}$ 26) $-\frac{8ac^3}{5b^7}$

4) $-\frac{1}{x^7}$ 11) $-\frac{17x^4}{y^6}$ 19) $-11x^5$ 27) $-x^5$

5) $\frac{11}{x}$ 12) $-25x^5$ 20) $-\frac{bc^6}{2}$ 28) $49x^{10}$

6) $-\frac{8}{x^3}$ 13) $-13xa^7$ 21) $12a^5 b^4$ 29) x^{36}

7) $-\frac{12}{x^5}$ 14) 81 22) $-\frac{p^7}{4n^4}$

   15) $\frac{16}{9}$ 23) $-\frac{8ac^5}{3b^6}$

Scientific Notation

1) 2.23×10^{-1} 4) 9×10^2 7) 3.3×10^1

2) 9×10^{-2} 5) 2×10^3 8) 9.4×10^3

3) 4.5×10^0 6) 6×10^{-3} 9) 1.47×10^3

TASC Subject Test – Mathematics

10) 5.2×10^4
11) 8×10^6
12) 9×10^{-5}
13) 2.158×10^6
14) 3.9×10^{-3}
15) 7.5×10^{-5}
16) 4.3×10^6

17) 1.3×10^5
18) 4×10^9
19) 9×10^{-5}
20) 3.9×10^{-3}
21) 0.4
22) 0.0012
23) 270,000

24) 0.0006
25) 0.0036
26) 550,000
27) 32,000
28) 3,880,000
29) 0.000007
30) 0.00000042

Square Roots

1) 4
2) 5
3) 1
4) 8
5) 0
6) 14
7) 2
8) 16
9) 6

10) 17
11) 13
12) 12
13) 10
14) 40
15) 50
16) 18
17) 23
18) $2\sqrt{5}$

19) 25
20) $3\sqrt{2}$
21) $5\sqrt{2}$
22) 32
23) $4\sqrt{10}$
24) $4\sqrt{2}$
25) 10
26) 42
27) 6

28) 13
29) 30
30) 6
31) $2\sqrt{13}$
32) $3\sqrt{10}$
33) $2\sqrt{7}$
34) 80
35) 120
36) $4\sqrt{3}$

Simplifying radical expressions

1) $x\sqrt{13}$
2) $5x\sqrt{3}$
3) $3\sqrt[3]{a}$
4) $8x^2\sqrt{x}$
5) $6\sqrt{6a}$
6) $w\sqrt[3]{63}$
7) $8\sqrt{3x}$
8) $5\sqrt{5v}$
9) $4\sqrt[3]{2x^2}$
10) $10x^4\sqrt{x}$
11) $4x^2$

12) $5a\sqrt[3]{4a^2}$
13) $11\sqrt{2}$
14) $14p\sqrt{2p}$
15) $2m^3\sqrt{2}$
16) $3x.y\sqrt{22xy}$
17) $11x^2y^2\sqrt{xy}$
18) $4a^3b\sqrt{b}$
19) $3x^2y^3\sqrt{10xy}$
20) $4x^2\sqrt[3]{y^2}$
21) $40x^2$
22) $54x$

23) $2y^2\sqrt[3]{7x^2}$
24) $10xy^2\sqrt[3]{x^2y}$
25) $40\sqrt{2a}$
26) $5x^2\sqrt[4]{y}$
27) $2x^2y^2r\sqrt{6yr}$
28) $30x^2y^2z^4\sqrt{y}$
29) $21x^3y^2\sqrt[3]{y}$
30) $45a^2bc^4\sqrt{ac}$
31) $5x^2y^4$

TASC Subject Test – Mathematics

Chapter 5 :
Algebraic Expressions

Topics that you'll practice in this chapter:

- ✓ Simplifying Variable Expressions
- ✓ Simplifying Polynomial Expressions
- ✓ Translate Phrases into an Algebraic Statement
- ✓ The Distributive Property
- ✓ Evaluating One Variable Expressions
- ✓ Evaluating Two Variables Expressions
- ✓ Combining like Terms

Mathematics is, as it were, a sensuous logic, and relates to philosophy as do the arts, music, and plastic art to poetry. — K. Shegel

TASC Subject Test – Mathematics

Simplifying Variable Expressions

✎ **Simplify each expression.**

1) $3(x + 5) =$

2) $(-4)(7x - 5) =$

3) $11x + 5 - 6x =$

4) $-4 - 2x^2 - 6x^2 =$

5) $7 + 13x^2 + 3 =$

6) $3x^2 + 7x + 15x^2 =$

7) $3x^2 - 12x^2 + 4x =$

8) $4x^2 - 8x - 2x =$

9) $6x + 7(3 - 4x) =$

10) $8x + 4(15x - 3) =$

11) $6(-3x - 9) - 17 =$

12) $-11x^2 - (-5x) =$

13) $2x + 7 + 5 - 8x =$

14) $7 + 6x - 11 - 5x =$

15) $27x + 8 - 13 - 5x =$

16) $(-11)(-5x + 2) - 41x =$

17) $19x - 4(4 - 2x) =$

18) $16x + 3(3x + 6) + 10 =$

19) $5(-2x - 4) - 13x =$

20) $16x - 3x(x + 10) =$

21) $17x + 5x(2 - 4x) =$

22) $5x(-4x - 7) + 20x =$

23) $25x - 19 + 4x^2 =$

24) $6x(x - 11) + 25 =$

25) $4x - 5 + 15x + 3x^2 =$

26) $-7x^2 - 11x - 9x =$

27) $10x - 9x^2 - 3x^2 - 7 =$

28) $13 + 3x^2 - 9x^2 - 21x =$

29) $22x + 10x^2 - 15x + 17 =$

30) $4x^2 + 25x + 21x^2 =$

31) $29 - 12x^2 - 23x - 4x^2 =$

32) $22x - 19x - 9x^2 + 30 =$

WWW.MathNotion.Com

TASC Subject Test – Mathematics

Simplifying Polynomial Expressions

✏️ Simplify each polynomial.

1) $(2x^3 + 8x^2) - (11x + 3x^2) =$ _____

2) $(2x^5 + 7x^3) - (5x^3 + 11x^2) =$ _____

3) $(41x^4 + 5x^2) - (4x^2 + 20x^4) =$ _____

4) $13x - 8x^2 + 4(4x^2 + 3x^3) =$ _____

5) $(4x^3 - 22) + 5(3x^2 - 6x^3) =$ _____

6) $(4x^3 - 3x) - 5(2x^3 + x^4) =$ _____

7) $5(5x - 2x^3) - 2(8x^3 + 5x^2) =$ _____

8) $(3x^2 - 10x) - (5x^3 + 14x^2) =$ _____

9) $5x^3 - (3x^4 + 5x) + 2x^2 =$ _____

10) $11x^4 - (3x^2 + 5x) + 7x =$ _____

11) $(6x^2 - 3x^4) - (10x^4 + 3x^2) =$ _____

12) $2x^2 - 7x^3 + 19x^4 - 22x^3 =$ _____

13) $10x^2 - x^4 + 4x^4 - 32x^3 =$ _____

14) $-5x^2 + 17x^3 - 8x^2 - 6x =$ _____

15) $x^4 - 11x^5 - 30x^4 + 5x^2 =$ _____

16) $21x^3 + 13x - 5x^2 - 11x^3 =$ _____

WWW.MathNotion.Com

TASC Subject Test – Mathematics

Translate Phrases into an Algebraic Statement

✍ **Write an algebraic expression for each phrase.**

1) 9 multiplied by x. _____

2) Subtract 11 from y. _____

3) 19 divided by x. _____

4) 38 decreased by y. _____

5) Add y to 40. _____

6) The square of 6. _____

7) x raised to the fifth power. _____

8) The sum of six and a number. _____

9) The difference between fifty-seven and y. _____

10) The quotient of nine and a number. _____

11) The quotient of the square of x and 25. _____

12) The difference between x and 6 is 19. _____

13) 10 times a reduced by the square of b. _____

14) Subtract the product of a and b from 41. _____

TASC Subject Test – Mathematics

The Distributive Property

✎ Use the distributive property to simply each expression.

1) $4(1 + 2x) =$

2) $2(4 + 7x) =$

3) $3(4x - 4) =$

4) $(2x - 5)(-6) =$

5) $(-3)(x + 6) =$

6) $(4 + 3x)2 =$

7) $(-5)(8 - 3x) =$

8) $-(-5 - 7x) =$

9) $(-6x + 3)(-3) =$

10) $(-4)(x - 7) =$

11) $-(5 - 3x) =$

12) $3(9 + 4x) =$

13) $6(4 + 3x) =$

14) $(-5x + 3)2 =$

15) $(5 - 8x)(-3) =$

16) $(-12)(3x + 3) =$

17) $(5 - 3x)6 =$

18) $4(2 + 6x) =$

19) $8(7x - 3) =$

20) $(-2x + 3)4 =$

21) $(7 - 5x)(-9) =$

22) $(-10)(x - 8) =$

23) $(11 - 4x)3 =$

24) $(-6)(10x - 4) =$

25) $(3 - 9x)(-7) =$

26) $(-9)(x + 9) =$

27) $(-3 + 5x)(-7) =$

28) $(-5)(8 - 10x) =$

29) $12(4x - 8) =$

30) $(-10x + 13)(-3) =$

31) $(-8)(3x - 2) + 4(x + 5) =$

32) $(-8)(x + 4) - (6 + 5x) =$

WWW.MathNotion.Com

TASC Subject Test – Mathematics

Evaluating One Variable Expressions

✎ **Evaluate each expression using the value given.**

1) $8 - x$, $x = 5$

2) $x - 9$, $x = 5$

3) $5x + 4$, $x = 3$

4) $x - 13$, $x = -4$

5) $12 - x$, $x = 4$

6) $x + 2$, $x = 6$

7) $4x + 8$, $x = 3$

8) $x + (-7)$, $x = -8$

9) $4x + 5$, $x = 2$

10) $3x + 9$, $x = -2$

11) $15 + 3x - 7$, $x = 2$

12) $17 - 3x$, $x = 3$

13) $8x - 9$, $x = 4$

14) $5x + 4$, $x = -3$

15) $10x + 5$, $x = 3$

16) $14 - 4x$, $x = -6$

17) $3(5x + 3)$, $x = 9$

18) $4(-3x - 6)$, $x = 3$

19) $7x - 2x + 12$, $x = 4$

20) $(5x + 6) \div 2$, $x = 8$

21) $(x + 18) \div 10$, $x = 12$

22) $5x - 12 + 3x$, $x = -3$

23) $(6 - 4x)(-3)$, $x = -4$

24) $9x^2 + 3x - 6$, $x = 2$

25) $x^2 - 10x$, $x = -5$

26) $3x(7 - 2x)$, $x = 2$

27) $12x + 6 - 2x^2$, $x = -4$

28) $(-3)(4x - 8 + 3x)$, $x = 3$

29) $(-6) + \frac{x}{4} + 3x$, $x = 16$

30) $(-6) + \frac{x}{5}$, $x = 35$

31) $\left(-\frac{45}{x}\right) - 7 + 2x$, $x = 9$

32) $\left(-\frac{21}{x}\right) - 12 + 4x$, $x = 7$

WWW.MathNotion.Com

TASC Subject Test – Mathematics

Evaluating Two Variables Expressions

✎ **Evaluate each expression using the values given.**

1) $2x - 4y$,
   $x = 4, y = 1$

2) $3x + 5y$,
   $x = -2, y = 2$

3) $-7a + 4b$,
   $a = 2, b = 4$

4) $3x + 5 - y$,
   $x = 5, y = 6$

5) $3z + 12 - 2k$,
   $z = 5, k = 6$

6) $6(-x - 3y)$,
   $x = 5, y = -2$

7) $5a + 3b$,
   $a = 3, b = 4$

8) $7x \div 3y$,
   $x = 3, y = 7$

9) $2x + 15 + 5y$,
   $x = -3, y = 1$

10) $5a - (18 - b)$,
    $a = 2, b = 8$

11) $2z + 20 + 5k$,
    $z = -6, k = 5$

12) $xy + 10 + 4x$,
    $x = 3, y = 5$

13) $2x + 4y - 8 + 5$,
    $x = 5, y = 2$

14) $\left(-\frac{24}{x}\right) + 3 + 2y$,
    $x = 4, y = 6$

15) $(-3)(-3a - 3b)$,
    $a = 4, b = 5$

16) $12 + 4x - 7 - y$,
    $x = 3, y = 5$

17) $11x + 5 - 8y + 6$,
    $x = 5, y = 2$

18) $10 + 2(-4x - 5y)$,
    $x = 5, y = 4$

19) $5x + 13 + 6y$,
    $x = 5, y = 6$

20) $10a - (7a + 3b) - 11$,
    $a = 3, b = 8$

WWW.MathNotion.Com

TASC Subject Test – Mathematics

Combining like Terms

✎ **Simplify each expression.**

1) $11x + 3x + 6 =$

2) $8(2x - 6) =$

3) $18x - 7x + 11 =$

4) $(-4)(6x - 7) =$

5) $22x - 10x - 5 =$

6) $32x - 13 + 8x =$

7) $15 - (8x - 11) =$

8) $-24x + 17 - 11x =$

9) $12x - 8 - 6x + 9 =$

10) $21x + 5 - 36 + 12x =$

11) $28x + 3x - 11 =$

12) $(-3x + 4)5 =$

13) $2 + 4x + 9x - 8 =$

14) $6(2x - 5x) - 4 =$

15) $4(5x + 11) + 3x =$

16) $x - 14 - 11x =$

17) $5(10 + 9x) - 8x =$

18) $42x + 17 - 23x =$

19) $(-7x) + 19 + 20x =$

20) $(-7x) - 33 + 29x =$

21) $4(5x + 3) - 19x =$

22) $5(6 - 2x) - 15x =$

23) $-24x + (11 - 18x) =$

24) $(-9) - (6)(7x + 3) =$

25) $(-1)(8x - 10) - 21x =$

26) $-36x + 14 + 27x - 5x =$

27) $3(-13x + 6) - 17x =$

28) $-5x - 42 + 32x =$

29) $37x - 19x + 15 - 9x =$

30) $3(5x + 7x) - 31 =$

31) $14 - 6x - 15 - 9x =$

32) $-2(-5x - 7x) + 27x =$

WWW.MathNotion.Com

TASC Subject Test – Mathematics

Answers of Worksheets

Simplifying Variable Expressions

1) $3x + 15$
2) $-28x + 20$
3) $5x + 5$
4) $-8x^2 - 4$
5) $13x^2 + 10$
6) $18x^2 + 7x$
7) $-9x^2 + 4x$
8) $4x^2 - 10x$
9) $-22x + 21$
10) $68x - 12$
11) $-18x - 71$
12) $-11x^2 + 5x$
13) $-6x + 12$
14) $x - 4$
15) $22x - 5$
16) $14x - 22$
17) $27x - 16$
18) $25x + 28$
19) $-23x - 20$
20) $-3x^2 - 14x$
21) $-20x^2 + 27x$
22) $-20x^2 - 15x$
23) $4x^2 + 25x - 19$
24) $6x^2 - 66x + 25$
25) $3x^2 + 19x - 5$
26) $-7x^2 - 20x$
27) $-12x^2 + 10x - 7$
28) $-6x^2 - 21x + 13$
29) $10x^2 + 7x + 17$
30) $25x^2 + 25x$
31) $-16x^2 - 23x + 29$
32) $-9x^2 + 3x + 30$

Simplifying Polynomial Expressions

1) $2x^3 + 5x^2 - 11x$
2) $2x^5 + 2x^3 - 11x^2$
3) $21x^4 + x^2$
4) $12x^3 + 8x^2 + 13x$
5) $-26x^3 + 15x^2 - 22$
6) $-5x^4 - 6x^3 - 3x$
7) $-26x^3 - 10x^2 + 25x$
8) $-5x^3 - 11x^2 - 10x$
9) $-3x^4 + 5x^3 + 2x^2 - 5x$
10) $11x^4 - 3x^2 + 2x$
11) $-13x^4 + 3x^2$
12) $19x^4 - 29x^3 + 2x^2$
13) $3x^4 - 32x^3 + 10x^2$
14) $17x^3 - 13x^2 - 6x$
15) $-11x^5 - 29x^4 + 5x^2$
16) $10x^3 - 5x^2 + 13x$

Translate Phrases into an Algebraic Statement

1) $9x$
2) $y - 11$
3) $\frac{19}{x}$
4) $38 - y$
5) $y + 40$
6) 6^2
7) x^5
8) $6 + x$
9) $57 - y$
10) $\frac{9}{x}$
11) $\frac{x^2}{25}$
12) $x - 6 = 19$
13) $10a - b^2$
14) $41 - ab$

The Distributive Property

1) $8x + 4$
2) $14x + 8$
3) $12x - 12$
4) $-12x + 30$
5) $-3x - 18$
6) $6x + 8$
7) $15x - 40$
8) $7x + 5$
9) $18x - 9$
10) $-4x + 28$
11) $3x - 5$
12) $12x + 27$

TASC Subject Test – Mathematics

13) $18x + 24$
14) $-10x + 6$
15) $24x - 15$
16) $-36x - 36$
17) $-18x + 30$

18) $24x + 8$
19) $56x - 24$
20) $-8x + 12$
21) $45x - 63$
22) $-10x + 80$

23) $-12x + 33$
24) $-60x + 24$
25) $63x - 21$
26) $-9x - 81$
27) $-35x + 21$

28) $50x - 40$
29) $48x - 96$
30) $30x - 39$
31) $-20x + 36$
32) $-13x - 38$

Evaluating One Variables

1) 3
2) -4
3) 19
4) -17
5) 8
6) 8
7) 20
8) -15

9) 13
10) 3
11) 14
12) 8
13) 23
14) -11
15) 35
16) 38

17) 144
18) -60
19) 32
20) 23
21) 3
22) -36
23) -66
24) 36

25) 75
26) 18
27) -74
28) -39
29) 46
30) 1
31) 6
32) 13

Evaluating Two Variables

1) 4
2) 4
3) 2
4) 14
5) 15

6) 6
7) 27
8) 1
9) 14
10) 0

11) 33
12) 37
13) 15
14) 9
15) 81

16) 12
17) 50
18) -70
19) 74
20) -26

Combining like Terms

1) $14x + 6$
2) $16x - 48$
3) $11x + 11$
4) $-24x + 28$
5) $12x - 5$
6) $40x - 13$
7) $-8x + 26$
8) $-35x + 17$

9) $6x + 1$
10) $33x - 31$
11) $31x - 11$
12) $-15x + 20$
13) $13x - 6$
14) $-18x - 4$
15) $23x + 44$
16) $-10x - 14$

17) $37x + 50$
18) $19x + 17$
19) $13x + 19$
20) $22x - 33$
21) $x + 12$
22) $-25x + 30$
23) $-42x + 11$
24) $-42x - 27$

25) $-29x + 10$
26) $-14x + 14$
27) $-56x + 18$
28) $27x - 42$
29) $9x + 15$
30) $36x - 31$
31) $-15x - 1$
32) $51x$

WWW.MathNotion.Com

TASC Subject Test – Mathematics

Chapter 6 : Equations and Inequalities

Topics that you'll practice in this chapter:

- ✓ One–Step Equations
- ✓ Multi-Step Equations
- ✓ Graphing Single-Variable Inequalities
- ✓ One-Step Inequalities
- ✓ Multi-Step Inequalities
- ✓ Systems of Equations
- ✓ Systems of Equations Word Problems

"Life is a math equation. In order to gain the most, you have to know how to convert negatives into positives." – Anonymous

TASC Subject Test – Mathematics

One–Step Equations

✏ **Find the answer for each equation.**

1) $3x = 90, x =$ ___

2) $5x = 35, x =$ ___

3) $6x = 24, x =$ ___

4) $24x = 144, x =$ ___

5) $x + 15 = 20, x =$ ___

6) $x - 7 = 4, x =$ ___

7) $x - 9 = 2, x =$ ___

8) $x + 15 = 23, x =$ ___

9) $x - 4 = 13, x =$ ___

10) $12 = 16 + x, x =$ ___

11) $x - 10 = 2, x =$ ___

12) $5 - x = -11, x =$ ___

13) $28 = -6 + x, x =$ ___

14) $x - 20 = -35, x =$ ___

15) $x + 14 = -4, x =$ ___

16) $14 = 28 - x, x =$ ___

17) $7 + x = -7, x =$ ___

18) $x - 16 = 4, x =$ ___

19) $30 = x - 15, x =$ ___

20) $x - 5 = -18, x =$ ___

21) $x - 10 = 24, x =$ ___

22) $x - 20 = -25, x =$ ___

23) $x - 17 = 30, x =$ ___

24) $-70 = x - 28, x =$ ___

25) $x - 9 = 13, x =$ ___

26) $36 = 4x, x =$ ___

27) $x - 35 = 25, x =$ ___

28) $x - 25 = 10, x =$ ___

29) $70 - x = 16, x =$ ___

30) $x - 10 = 14, x =$ ___

31) $17 - x = -13, x =$ ___

32) $x - 9 = -30, x =$ ___

WWW.MathNotion.Com

TASC Subject Test – Mathematics

Multi-Step Equations

✎ **Find the answer for each equation.**

1) $3x + 3 = 9$

2) $-x + 5 = 12$

3) $4x - 8 = 8$

4) $-(3 - x) = 5$

5) $4x - 8 = 16$

6) $12x - 15 = 9$

7) $2x - 18 = 2$

8) $4x + 8 = 16$

9) $24x + 27 = 75$

10) $-14(3 + x) = 14$

11) $-3(2 + x) = 6$

12) $12 = -(x - 7)$

13) $3(3 - x) = 30$

14) $-15 = -(3x + 6)$

15) $40(3 + x) = 40$

16) $5(x - 10) = 25$

17) $-18 = x + 8x$

18) $3x + 25 = -2x - 10$

19) $7(6 + 3x) = -63$

20) $18 - 3x = -4 - 5x$

21) $4 - 6x = 36 + 2x$

22) $15 + 15x = -5 + 5x$

23) $42 = (-6x) - 7 + 7$

24) $21 = 3x - 21 + 4x$

25) $-18 = -6x - 9 + 3x$

26) $5x - 15 = -29 + 6x$

27) $7x - 18 = 4x + 3$

28) $-7 - 4x = 5(4 - x)$

29) $x - 5 = -5(-3 - x)$

30) $13x - 68 = 15x - 102$

31) $-5x - 3 = -3(9 + 3x)$

32) $-2x - 15 = 6x + 17$

TASC Subject Test – Mathematics

Graphing Single–Variable Inequalities

✎ Draw a graph for each inequality.

1) $x > -1$

2) $x \leq 2$

3) $x \geq 0$

4) $x < -3$

5) $x < \frac{1}{2}$

6) $x \leq -2$

7) $x \leq 3$

8) $x \geq -\frac{7}{2}$

WWW.MathNotion.Com

TASC Subject Test – Mathematics

One–Step Inequalities

✏️ **Find the answer for each inequality and graph it.**

1) $x + 4 \geq 4$

2) $x - 5 \leq 2$

3) $5x > 35$

4) $9 + x \leq 11$

5) $x - 5 < -9$

6) $9x \geq 72$

7) $9x \leq 27$

8) $x + 19 > 16$

WWW.MathNotion.Com

TASC Subject Test – Mathematics

Multi-Step Inequalities

✏ Calculate each inequality.

1) $x - 3 \leq 7$

2) $8 - x \leq 8$

3) $3x - 9 \leq 9$

4) $4x - 4 \geq 8$

5) $x - 7 \geq 1$

6) $5x - 15 \leq 5$

7) $6x - 8 \leq 4$

8) $-11 + 6x \leq 12$

9) $4(x - 4) \leq 16$

10) $3x - 10 \leq 11$

11) $5x - 25 < 25$

12) $9x - 5 < 22$

13) $20 - 7x \geq -15$

14) $33 + 6x < 45$

15) $8 + 8x \geq 96$

16) $7 + 3x < 13$

17) $4x - 3 < 9$

18) $5(2 - 2x) \geq -30$

19) $-(7 + 6x) < 29$

20) $12 - 8x \geq -20$

21) $-4(x - 6) > 24$

22) $\dfrac{3x + 9}{6} \leq 10$

23) $\dfrac{4x - 10}{3} \leq 2$

24) $\dfrac{2x - 8}{3} > 2$

25) $8 + \dfrac{x}{6} < 9$

26) $\dfrac{9x}{7} - 4 < 5$

27) $\dfrac{15x + 45}{15} > 1$

28) $16 + \dfrac{x}{4} < 6$

WWW.MathNotion.Com

TASC Subject Test – Mathematics

Systems of Equations

✎ Calculate each system of equations.

1) $-x + y = 2$ $x = $ ___
    $-4x + 2y = 6$ $y = $ ___

2) $-15x + 3y = -9$ $x = $ ___
    $9x - 16y = 48$ $y = $ ___

3) $y = -7$ $x = $ ___
    $6x + 5y = 7$ $y = $ ___

4) $3y = -9x + 15$ $x = $ ___
    $5x - 4y = -3$ $y = $ ___

5) $10x - 9y = -13$ $x = $ ___
    $-5x + 3y = 11$ $y = $ ___

6) $-12x - 16y = 20$ $x = $ ___
    $6x - 12y = 30$ $y = $ ___

7) $5x - 14y = -23$ $x = $ ___
    $-18x + 21y = 24$ $y = $ ___

8) $15x - 21y = -6$ $x = $ ___
    $2x - 3y = -2$ $y = $ ___

9) $-x + 3y = 3$ $x = $ ___
    $-14x + 16y = -10$ $y = $ ___

10) $x + 5y = 50$ $x = $ ___
    $3x + 10y = 80$ $y = $ ___

11) $6x - 7y = -8$ $x = $ ___
    $-x - 4y = -9$ $y = $ ___

12) $2x + 4y = -10$ $x = $ ___
    $2x - 8y = 14$ $y = $ ___

13) $4x + 3y = 12$ $x = $ ___
    $5x - 3y = 15$ $y = $ ___

14) $3x - 2y = 3$ $x = $ ___
    $7x - 8y = 22$ $y = $ ___

15) $3x + 2y = 5$ $x = $ ___
    $-10x - 4y = -14$ $y = $ ___

16) $10x + 7y = 1$ $x = $ ___
    $-5x - 7y = 24$ $y = $ ___

WWW.MathNotion.Com

TASC Subject Test – Mathematics

Systems of Equations Word Problems

✎ **Find the answer for each word problem.**

1) Tickets to a movie cost $4 for adults and $3 for students. A group of friends purchased 8 tickets for $31.00. How many adults ticket did they buy? ___

2) At a store, Eva bought two shirts and five hats for $77.00. Nicole bought three same shirts and four same hats for $84.00. What is the price of each shirt? ____

3) A farmhouse shelters 18 animals, some are pigs, and some are ducks. Altogether there are 66 legs. How many pigs are there? ____

4) A class of 214 students went on a field trip. They took 36 vehicles, some cars and some buses. If each car holds 5 students and each bus hold 22 students, how many buses did they take? ____

5) A theater is selling tickets for a performance. Mr. Smith purchased 5 senior tickets and 3 child tickets for $105 for his friends and family. Mr. Jackson purchased 3 senior tickets and 5 child tickets for $79. What is the price of a senior ticket? $____

6) The difference of two numbers is 10. Their sum is 20. What is the bigger number? $____

7) The sum of the digits of a certain two–digit number is 7. Reversing its digits increase the number by 9. What is the number? ____

8) The difference of two numbers is 11. Their sum is 25. What are the numbers? _____

9) The length of a rectangle is 5 meters greater than 2 times the width. The perimeter of rectangle is 28 meters. What is the length of the rectangle? _____

10) Jim has 25 nickels and dimes totaling $1.80. How many nickels does he have? ____

WWW.MathNotion.Com

TASC Subject Test – Mathematics

Answers of Worksheets

One–Step Equations

1) 30	9) 17	17) −14	25) 22
2) 7	10) −4	18) 20	26) 9
3) 4	11) 12	19) 45	27) 60
4) 6	12) 16	20) −13	28) 35
5) 5	13) 34	21) 34	29) 54
6) 11	14) −15	22) −5	30) 24
7) 11	15) −18	23) 47	31) 30
8) 8	16) 14	24) −42	32) −21

Multi–Step Equations

1) 2	9) 2	17) −2	25) 3
2) −7	10) −4	18) −7	26) 14
3) 4	11) −4	19) −5	27) 7
4) 8	12) −5	20) −11	28) 27
5) 6	13) −7	21) −4	29) −5
6) 2	14) 3	22) −2	30) 17
7) 10	15) −2	23) −7	31) −6
8) 2	16) 15	24) 6	32) −4

Graphing Single–Variable Inequalities

1) open circle at −1, shaded right (x > −1)

2) closed circle at 2, shaded left (x ≤ 2)

3) open circle at 0, shaded right (x > 0)

4) open circle at −3, shaded left (x < −3)

WWW.MathNotion.Com

TASC Subject Test – Mathematics

5)
6)
7)
8)

One–Step Inequalities

1)
2)
3)
4)
5)
6)
7)
8)

Multi-Step Inequalities

1) $x \leq 10$
2) $x \geq 0$
3) $x \leq 6$
4) $x \geq 3$
5) $x \geq 8$
6) $x \leq 4$
7) $x \leq 2$
8) $x \leq \frac{23}{6}$
9) $x \leq 8$
10) $x \leq 7$
11) $x < 10$
12) $x < 3$
13) $x \leq 5$
14) $x < 2$
15) $x \geq 11$
16) $x < 2$
17) $x < 3$
18) $x \leq 4$
19) $x > -6$
20) $x \leq 4$
21) $x < 0$
22) $x \leq 17$
23) $x \leq 4$

TASC Subject Test – Mathematics

24) $x > 7$
25) $x < 6$
26) $x < 7$
27) $x > -2$
28) $x < -40$

Systems of Equations

1) $x = -1, y = 1$
2) $x = 0, y = -3$
3) $x = 7$
4) $x = 1, y = 2$
5) $x = -4, y = -3$
6) $x = 1, y = -2$
7) $x = 1, y = 2$
8) $x = 8, y = 6$
9) $x = 3, y = 2$
10) $x = -20, y = 14$
11) $x = 1, y = 2$
12) $x = -1, y = -2$
13) $x = 3, y = 0$
14) $x = -2, y = -\frac{9}{2}$
15) $x = 1, y = 1$
16) $x = 5, y = -7$

Systems of Equations Word Problems

1) 7
2) $16
3) 15
4) 2
5) $18
6) 15
7) 34
8) 18, 7
9) 11 meters
10) 14

WWW.MathNotion.Com

TASC Subject Test – Mathematics

TASC Subject Test – Mathematics

Chapter 7:
Linear Functions

Topics that you'll practice in this chapter:

- ✓ Finding Slope
- ✓ Graphing Lines Using Line Equation
- ✓ Writing Linear Equations
- ✓ Graphing Linear Inequalities
- ✓ Finding Midpoint
- ✓ Finding Distance of Two Points

"Nature is written in mathematical language." – Galileo Galilei

TASC Subject Test – Mathematics

Finding Slope

✍ **Find the slope of each line.**

1) $y = x + 8$

2) $y = -3x + 5$

3) $y = 2x + 12$

4) $y = -4x + 19$

5) $y = 11 + 6x$

6) $y = 7 - 5x$

7) $y = 8x + 19$

8) $y = -9x + 20$

9) $y = -7x + 4$

10) $y = 3x - 8$

11) $y = \frac{1}{3}x + 8$

12) $y = -\frac{4}{5}x + 9$

13) $-3x + 6y = 30$

14) $4x + 4y = 16$

15) $3y - x = 10$

16) $8y - x = 5$

✍ **Find the slope of the line through each pair of points.**

17) $(2, 3), (7, 10)$

18) $(-3, 5), (2, 15)$

19) $(5, -3), (1, 9)$

20) $(-5, -5), (10, 25)$

21) $(22, 3), (7, 18)$

22) $(-16, 8), (-7, 26)$

23) $(25, 11), (29, 19)$

24) $(26, -19), (14, 17)$

25) $(22, -13), (20, -11)$

26) $(19, 7), (15, -3)$

27) $(5, 7), (11, 19)$

28) $(52, -62), (40, 70)$

WWW.MathNotion.Com

TASC Subject Test – Mathematics

Graphing Lines Using Line Equation

✎ Sketch the graph of each line.

1) $y = x - 2$

2) $y = -3x + 2$

3) $x + y = 0$

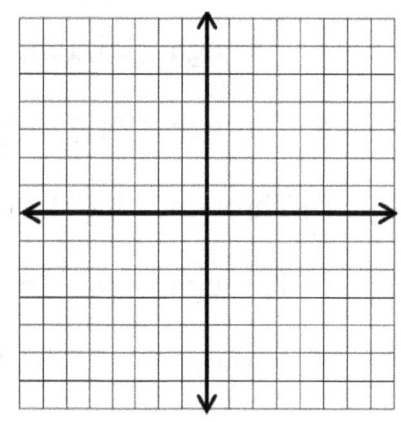

4) $x + y = -3$

5) $2x + 3y = -4$

6) $y - 3x + 6 = 0$

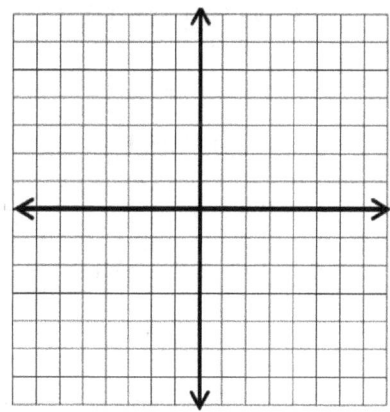

TASC Subject Test – Mathematics

Writing Linear Equations

✎ **Write the equation of the line through the given points.**

1) Through: $(2, -5), (3, 9)$

2) Through: $(-6, 3), (3, 12)$

3) Through: $(10, 7), (5, 27)$

4) Through: $(15, 11), (3, -1)$

5) Through: $(24, 17), (12, -7)$

6) Through: $(8, 29), (4, -7)$

7) Through: $(20, -16), (12, 0)$

8) Through: $(-3, 10), (2, -5)$

9) Through: $(-6, 17), (4, -3)$

10) Through: $(-8, 22), (5, -4)$

11) Through: $(9, 27), (3, -3)$

12) Through: $(11, 32), (9, 4)$

13) Through: $(-3, 13), (-4, 0)$

14) Through: $(-5, 5), (5, 15)$

15) Through: $(18, -32), (11, 3)$

16) Through: $(-4, 25), (4, -15)$

✎ **Find the answer for each problem.**

17) What is the equation of a line with slope 6 and intercept 12?
_____

18) What is the equation of a line with slope -11 and intercept -4?
_____

19) What is the equation of a line with slope -3 and passes through point $(5, 2)$? _____

20) What is the equation of a line with slope -5 and passes through point $(-2, -1)$? _____

21) The slope of a line is -10 and it passes through point $(-3, 0)$. What is the equation of the line? _____

22) The slope of a line is 8 and it passes through point $(0, 7)$. What is the equation of the line? _____

TASC Subject Test – Mathematics

Graphing Linear Inequalities

✎ Sketch the graph of each linear inequality.

1) $y > 4x - 5$ 2) $y < 2x + 4$ 3) $y \leq -5x - 2$

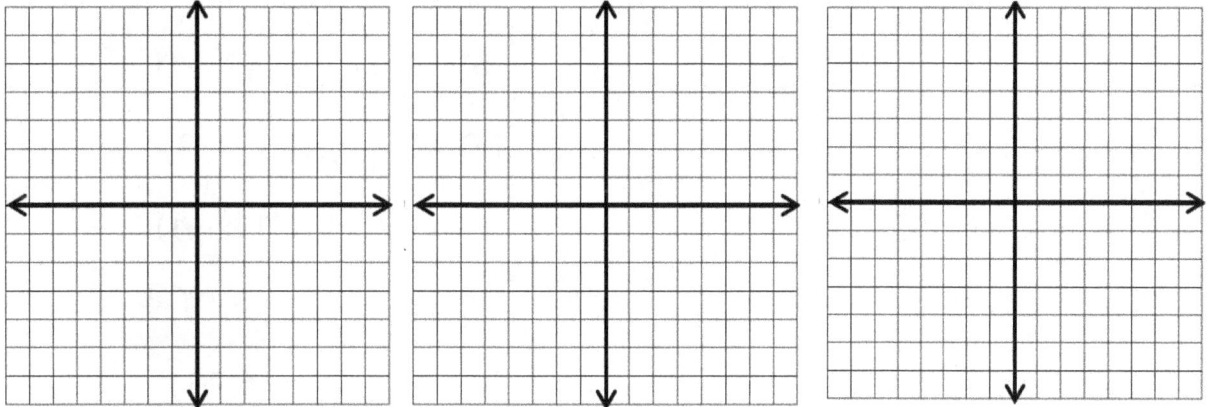

4) $4y \geq 12 + 4x$ 5) $-12y < 3x - 24$ 6) $5y \geq -15x + 10$

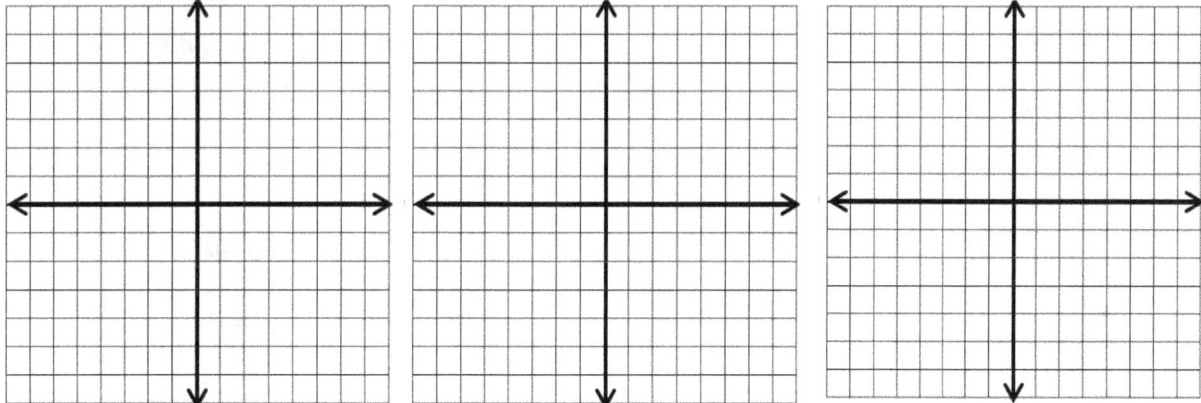

WWW.MathNotion.Com

TASC Subject Test – Mathematics

Finding Midpoint

✎ **Find the midpoint of the line segment with the given endpoints.**

1) $(-4, -3), (2, 3)$

2) $(9, 0), (-1, 8)$

3) $(9, -6), (3, 14)$

4) $(-10, -6), (0, 8)$

5) $(2, -5), (14, -15)$

6) $(-10, -3), (4, -13)$

7) $(8, 7), (-8, 13)$

8) $(-3, 6), (-9, 2)$

9) $(-4, 5), (16, -9)$

10) $(7, 14), (9, -2)$

11) $(-8, 6), (6, 6)$

12) $(10, 5), (-2, -3)$

13) $(-5, 12), (-3, 3)$

14) $(12, 7), (8, -2)$

15) $(10, 2), (-6, 14)$

16) $(-1, -2), (-7, 10)$

17) $(7, -7), (13, -13)$

18) $(-3, -8), (11, -4)$

19) $(5, -11), (-8, 9)$

20) $(14, -4), (16, 14)$

21) $(0, -5), (8, -1)$

22) $(3, 0), (-21, 18)$

23) $(17, -3), (-7, -5)$

24) $(26, -12), (6, 24)$

✎ **Find the answer for each problem.**

25) One endpoint of a line segment is $(-3, 7)$ and the midpoint of the line segment is $(-6, 9)$. What is the other endpoint? _____

26) One endpoint of a line segment is $(-3, 7)$ and the midpoint of the line segment is $(1, 5)$. What is the other endpoint? _____

27) One endpoint of a line segment is $(-10, -16)$ and the midpoint of the line segment is $(2, 9)$. What is the other endpoint? _____

WWW.MathNotion.Com

TASC Subject Test – Mathematics

Finding Distance of Two Points

✎ Find the distance between each pair of points.

1) $(6, 3), (-3, -9)$

2) $(5, 2), (-10, -6)$

3) $(8, 5), (8, 3)$

4) $(-8, -2), (2, 22)$

5) $(6, -7), (-3, -7)$

6) $(12, 0), (-9, -20)$

7) $(3, 20), (3, -5)$

8) $(10, 17), (5, 5)$

9) $(7, -2), (-4, -2)$

10) $(13, 4), (5, -2)$

11) $(11, 13), (5, 5)$

12) $(1, 4), (-23, -3)$

13) $(9, 8), (5, -4)$

14) $(-11, -4), (5, 8)$

15) $(-2, -6), (-2, -12)$

16) $(-1, -4), (23, 3)$

17) $(19, 3), (7, -6)$

18) $(-5, -2), (3, 4)$

19) $(2, 6), (2, -12)$

20) $(-4, -2), (8, -2)$

✎ Find the answer for each problem.

21) Triangle ABC is a right triangle on the coordinate system and its vertices are $(-2, 5)$, $(-2, 1)$, and $(1, 1)$. What is the area of triangle ABC? _____

22) Three vertices of a triangle on a coordinate system are $(3, -6)$, $(-5, -12)$, and $(3, -18)$. What is the perimeter of the triangle? _____

23) Four vertices of a rectangle on a coordinate system are $(-2, 2), (-2, 6)$, $(4, 2)$, and $(4, 6)$. What is its perimeter? _____

WWW.MathNotion.Com

TASC Subject Test – Mathematics

Answers of Worksheets

Finding Slope

1) 1
2) -3
3) 2
4) -4
5) 6
6) -5
7) 8
8) -9
9) -7
10) 3
11) $\frac{1}{3}$
12) $-\frac{4}{5}$
13) $\frac{1}{2}$
14) -1
15) $\frac{1}{3}$
16) $\frac{1}{8}$
17) $\frac{7}{5}$
18) 2
19) -3
20) 2
21) -1
22) 2
23) 2
24) -3
25) -1
26) $\frac{5}{2}$
27) 2
28) -11

Graphing Lines Using Line Equation

1) $y = x - 2$

2) $y = -3x + 2$

3) $x + y = 0$

4) $x + y = -3$

5) $2x + 3y = -4$

6) $y - 3x + 6 = 0$

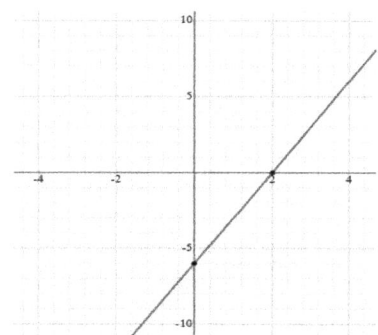

TASC Subject Test – Mathematics

Writing Linear Equations

1) $y = 14x - 33$
2) $y = x + 9$
3) $y = -4x + 47$
4) $y = x - 4$
5) $y = 2x - 31$
6) $y = 9x - 43$
7) $y = -2x + 24$
8) $y = -3x + 1$
9) $y = -2x + 5$
10) $y = -2x + 6$
11) $y = 5x - 18$
12) $y = 14x - 122$
13) $y = 13x + 52$
14) $y = x + 10$
15) $y = -5x + 58$
16) $y = -5x + 5$
17) $y = 6x + 12$
18) $y = -11x - 4$
19) $y = -3x + 17$
20) $y = -5x - 11$
21) $y = -10x - 30$
22) $y = 8x + 7$

Graphing Linear Inequalities

1) $y > 4x - 5$

2) $y < 2x + 4$

3) $y \leq -5x - 2$

4) $4y \geq 12 + 4x$

5) $-12y < 3x - 24$

6) $5y \geq -15x + 10$

Finding Midpoint

1) $(-1, 0)$
2) $(4, 4)$
3) $(6, 4)$
4) $(-5, 1)$
5) $(8, -10)$
6) $(-3, -8)$
7) $(0, 10)$
8) $(-6, 4)$
9) $(6, -2)$
10) $(8, 6)$
11) $(-1, 6)$
12) $(4, 1)$
13) $(-4, 7.5)$
14) $(10, 2.5)$
15) $(2, 8)$
16) $(-4, 4)$
17) $(10, -10)$
18) $(4, -6)$

WWW.MathNotion.Com

TASC Subject Test – Mathematics

19) $(-1.5, -1)$
20) $(15, 5)$
21) $(4, -3)$
22) $(-9, 9)$
23) $(5, -4)$
24) $(16, 6)$
25) $(-9, 11)$
26) $(5, 3)$
27) $(14, 34)$

Finding Distance of Two Points

1) 15
2) 17
3) 2
4) 26
5) 9
6) 29
7) 25
8) 13
9) 11
10) 10
11) 10
12) 25
13) $4\sqrt{10}$
14) 20
15) 6
16) 25
17) 15
18) 10
19) 18
20) 12
21) 6 *square units*
22) 32 *units*
23) 20 *units*

WWW.MathNotion.Com

TASC Subject Test – Mathematics

Chapter 8 : Polynomials

Topics that you'll practice in this chapter:

- ✓ Writing Polynomials in Standard Form
- ✓ Simplifying Polynomials
- ✓ Adding and Subtracting Polynomials
- ✓ Multiplying Monomials
- ✓ Multiplying and Dividing Monomials
- ✓ Multiplying a Polynomial and a Monomial
- ✓ Multiplying Binomials
- ✓ Factoring Trinomials
- ✓ Operations with Polynomials

Mathematics is the supreme judge; from its decisions there is no appeal. – Tobias Dantzig

TASC Subject Test – Mathematics

Writing Polynomials in Standard Form

✎ Write each polynomial in standard form.

1) $11x - 7x =$

2) $-5 + 19x - 19x =$

3) $6x^5 - 12x^3 =$

4) $12 + 17x^4 - 12 =$

5) $5x^2 + 4x - 9x^3 =$

6) $-3x^2 + 12x^5 =$

7) $5x + 8x^3 - 2x^8 =$

8) $-7x^3 + 4x - 9x^6 =$

9) $3x^2 + 22 - 6x =$

10) $3 - 4x + 9x^4 =$

11) $13x^2 + 28x - 8x^3 =$

12) $16 + 4x^2 - 2x^3 =$

13) $19x^2 - 9x + 9x^4 =$

14) $3x^4 - 7x^2 - 2x^3 =$

15) $-51 + 3x^2 - 8x^4 =$

16) $7x^2 - 8x^6 + 4x^4 - 15 =$

17) $6x^4 - 4x^5 + 16 - 3x^3 =$

18) $-2x^6 + 4x - 7x^2 - 5x =$

19) $11x^7 + 8x^5 - 5x^7 - 3x^2 =$

20) $2x^2 - 12x^5 + 8x^2 + 3x^6 =$

21) $4x^5 - 11x^7 - 6x^3 + 16x^5 =$

22) $6x^3 + 3x^5 + 34x^4 - 8x^5 =$

23) $3x(4x + 5 - 2x^2) =$

24) $12x(x^6 + 4x^3) =$

25) $5x(3x^2 + 6x + 4) =$

26) $7x(4 - 2x + 6x^5) =$

27) $3x(4x^4 - 4x^3 + 2) =$

28) $4x(2x^5 + 6x^2 - 3) =$

29) $5x(3x^4 + 4x^3 + 2x) =$

30) $2x(3x - 2x^3 + 4x^6) =$

WWW.MathNotion.Com

TASC Subject Test – Mathematics

Simplifying Polynomials

✏️ **Simplify each expression.**

1) $3(4x - 20) =$

2) $5x(3x - 4) =$

3) $6x(5x - 7) =$

4) $3x(7x + 5) =$

5) $5x(4x - 3) =$

6) $6x(8x + 2) =$

7) $(3x - 2)(x - 4) =$

8) $(x - 5)(2x + 6) =$

9) $(x - 3)(x - 7) =$

10) $(3x + 4)(3x - 4) =$

11) $(5x - 4)(5x - 2) =$

12) $6x^2 + 6x^2 - 8x^4 =$

13) $3x - 2x^2 + 5x^3 + 7 =$

14) $7x + 4x^2 - 10x^3 =$

15) $12x^2 + 5x^5 - 6x^3 =$

16) $-5x^2 + 4x^6 + 6x^8 =$

17) $-12x^3 + 10x^5 - 4x^6 + 4x =$

18) $11 - 7x^2 + 4x^2 - 16x^3 + 11 =$

19) $2x^2 - 9x + 4x^3 + 15x - 10x =$

20) $13 - 7x^5 + 6x^5 - 4x^2 + 5 =$

21) $-5x^8 + x^6 - 14x^3 + 5x^8 =$

22) $(7x^4 - 4) + (7x^4 - 2x^4) =$

23) $3(3x^4 - 4x^3 - 6x^4) =$

24) $-5(x^9 + 8) - 5(10 - x^9) =$

25) $8x^3 - 9x^4 - 2x + 19 - 8x^3 =$

26) $11 - 8x^3 + 6x^3 - 7x^5 + 6 =$

27) $(5x^3 - 4x) - (6x - 2 - 6x^3) =$

28) $4x^2 - 5x^4 - x(3x^3 + 2x) =$

29) $6x + 6x^5 - 10 - 4(x^5 - 3) =$

30) $4 - 3x^4 + (6x^5 - 2x^4 + 5x^5) =$

31) $-(x^5 + 4) - 8(3 + x^5) =$

32) $(4x^3 - 3x) - (3x - 5x^3) =$

TASC Subject Test – Mathematics

Adding and Subtracting Polynomials

✎ **Add or subtract expressions.**

1) $(-2x^2 - 3) + (3x^2 + 4) =$

2) $(4x^3 + 6) - (7 - 2x^3) =$

3) $(4x^5 + 5x^2) - (2x^5 + 15) =$

4) $(6x^3 - 2x^2) + (5x^2 - 4x) =$

5) $(10x^4 + 28x) - (34x^4 + 6) =$

6) $(7x^2 - 3) + (7x^2 + 3) =$

7) $(9x^2 + 4) - (10 - 5x^2) =$

8) $(6x^2 + x^5) - (x^5 + 4) =$

9) $(4x^3 - x) + (3x - 7x^3) =$

10) $(11x + 10) - (8x + 10) =$

11) $(15x^3 - 3x) - (3x - 4x^3) =$

12) $(4x - x^5) - (6x^5 + 8x) =$

13) $(2x^2 - 7x^7) - (4x^7 - 6x) =$

14) $(3x^2 - 5) + (8x^2 + 4x^5) =$

15) $(9x^4 + 5x^5) - (x^5 - 9x^4) =$

16) $(-4x^3 - 2x) + (9x - 5x^3) =$

17) $(4x - 3x^2) - (148x^2 + x) =$

18) $(5x - 8x^4) - (3x^4 - 4x^2) =$

19) $(8x^4 - 4) + (2x^4 - 3x^2) =$

20) $(5x^6 + 7x^3) - (x^3 - 5x^6) =$

21) $(-2x^2 + 20x^5 + 5x^4) + (12x^4 + 8x^5 + 24x^2) =$

22) $(7x^4 - 9x^7 - 6x) - (-3x^4 - 9x^7 + 6x) =$

23) $(14x + 12x^4 - 18x^6) + (20x^4 + 18x^6 - 10x) =$

24) $(5x^8 - 6x^6 - 4x) - (5x^3 + 9x^6 - 7x) =$

25) $(11x^2 - 6x^4 - 3x) - (-4x^2 - 12x^4 + 9x) =$

26) $(-5x^9 + 14x^3 + 3x^7) + (10x^7 + 26x^3 + 3x^9) =$

WWW.MathNotion.Com

TASC Subject Test – Mathematics

Multiplying Monomials

✏️ **Simplify each expression.**

1) $6u^8 \times (-u^2) =$

2) $(-5p^8) \times (-2p^3) =$

3) $4xy^3z^5 \times 3z^4 =$

4) $3u^5t \times 8ut^4 =$

5) $(-5a^2) \times (-7a^3b^6) =$

6) $-3a^4b^3 \times 6a^2b =$

7) $13xy^5 \times x^4y^4 =$

8) $6p^4q^3 \times (-8pq^6) =$

9) $8s^4t^3 \times 4st^3 =$

10) $(-6x^4y^3) \times 6x^2y =$

11) $3xy^7z \times 12z^3 =$

12) $24xy \times x^2y =$

13) $13pq^4 \times (-3p^2q) =$

14) $13s^3t^4 \times st^4 =$

15) $11p^5 \times (-6p^3) =$

16) $(-8p^3q^5r) \times 3pq^4r^6 =$

17) $(-4a^4) \times (-7a^3b) =$

18) $6u^6v^2 \times (-5u^3v^4) =$

19) $9u^5 \times (-3u) =$

20) $-6xy^5 \times 4x^2y =$

21) $13y^5z^3 \times (-y^3z) =$

22) $8a^4bc^3 \times 2abc^3 =$

23) $(-7p^5q^6) \times (-5p^4q^2) =$

24) $4u^5v^3 \times (-4u^7v^3) =$

25) $17y^4z^5 \times (-y^6z) =$

26) $(-5pq^3r^2) \times 8p^2q^4r =$

27) $3ab^5c^6 \times 5a^4bc^2 =$

28) $6x^3yz^2 \times 3x^2y^7z^3 =$

WWW.MathNotion.Com

TASC Subject Test – Mathematics

Multiplying and Dividing Monomials

✎ **Simplify each expression.**

1) $(5x^5)(2x^2) =$

2) $(4x^4)(6x^2) =$

3) $(3x^4)(7x^4) =$

4) $(5x^6)(4x^2) =$

5) $(12x^4)(3x^6) =$

6) $(4yx^8)(8y^4x^3) =$

7) $(14x^4y)(x^3y^5) =$

8) $(-5x^3y^4)(2x^3y^5) =$

9) $(-6x^4y^2)(-3x^3y^5) =$

10) $(5x^3y)(-5x^2y^3) =$

11) $(6x^4y^3)(4x^3y^4) =$

12) $(4x^3y^2)(5x^2y^4) =$

13) $(12x^3y^6)(4x^4y^{10}) =$

14) $(15x^3y^5)(3x^4y^6) =$

15) $(7x^2y^7)(8x^6y^7) =$

16) $(-3x^3y^8)(7x^9y^4) =$

17) $\dfrac{5x^6y^6}{xy^4} =$

18) $\dfrac{19x^7y^5}{19x^6y} =$

19) $\dfrac{56x^4y^4}{8xy} =$

20) $\dfrac{81x^5y^6}{9x^4y^5} =$

21) $\dfrac{36x^7y^6}{9x^2y^3} =$

22) $\dfrac{48x^9y^7}{4x^4y^6} =$

23) $\dfrac{88x^{18}y^{12}}{11x^8y^9} =$

24) $\dfrac{30x^7y^6}{6x^8y^3} =$

25) $\dfrac{150x^7y^6}{30x^4y^6} =$

26) $\dfrac{-42x^{18}y^{14}}{6x^4y^9} =$

27) $\dfrac{-36x^7y^8}{9x^5y^8} =$

WWW.MathNotion.Com

TASC Subject Test – Mathematics

Multiplying a Polynomial and a Monomial

✏ **Find each product.**

1) $x(2x + 4) =$

2) $6(4 - 2x) =$

3) $5x(4x + 2) =$

4) $x(-4x + 5) =$

5) $8x(2x - 2) =$

6) $6(2x - 4y) =$

7) $7x(5x - 5) =$

8) $3x(12x + 2y) =$

9) $4x(x + 6y) =$

10) $11x(3x + 4y) =$

11) $7x(3x + 2) =$

12) $10x(4x - 10y) =$

13) $9x(3x - 2y) =$

14) $7x(x - 4y + 6) =$

15) $8x(2x^2 + 5y^2) =$

16) $12x(2x + 3y) =$

17) $4(2x^4 - 4y^4) =$

18) $4x(-3x^2y + 4y) =$

19) $-4(5x^3 - 2xy + 4) =$

20) $4(x^2 - 5xy - 6) =$

21) $8x(2x^3 - 5xy + 2x) =$

22) $-6x(-2x^3 - 6x + 2xy) =$

23) $3(2x^2 + xy - 9y^2) =$

24) $4x(5x^3 - 3x + 7) =$

25) $6(3x^{22} - 2x - 5) =$

26) $x^2(-2x^3 + 4x + 3) =$

27) $x^2(4x^3 + 10 - 2x) =$

28) $4x^4(3x^3 - 2x + 5) =$

29) $2x^2(4x^4 - 5xy + 7y^3) =$

30) $5x^2(5x^4 - 3x + 9) =$

31) $7x^2(6x^2 + 3x - 6) =$

32) $4x(x^3 - 4xy + 2y^2) =$

WWW.MathNotion.Com

TASC Subject Test – Mathematics

Multiplying Binomials

✎ **Find each product.**

1) $(x + 3)(x + 6) =$

2) $(x - 4)(x + 3) =$

3) $(x - 3)(x - 8) =$

4) $(x + 8)(x + 9) =$

5) $(x - 2)(x - 12) =$

6) $(x + 5)(x + 5) =$

7) $(x - 6)(x + 7) =$

8) $(x - 8)(x - 3) =$

9) $(x + 7)(x + 12) =$

10) $(x - 4)(x + 8) =$

11) $(x + 8)(x + 8) =$

12) $(x + 2)(x + 7) =$

13) $(x - 6)(x + 6) =$

14) $(x - 5)(x + 5) =$

15) $(x + 11)(x + 11) =$

16) $(x + 6)(x + 9) =$

17) $(x - 2)(x + 2) =$

18) $(x - 4)(x + 7) =$

19) $(3x + 5)(x + 6) =$

20) $(5x - 6)(4x + 8) =$

21) $(x - 7)(3x + 7) =$

22) $(x - 9)(x - 4) =$

23) $(x - 12)(x + 2) =$

24) $(2x - 4)(5x + 4) =$

25) $(3x - 8)(x + 8) =$

26) $(7x - 2)(6x + 3) =$

27) $(4x + 5)(3x + 5) =$

28) $(7x - 4)(9x + 4) =$

29) $(x + 2)(2x - 8) =$

30) $(5x - 4)(5x + 4) =$

31) $(3x + 2)(3x - 7) =$

32) $(x^2 + 8)(x^2 - 8) =$

WWW.MathNotion.Com

TASC Subject Test – Mathematics

Factoring Trinomials

✎ **Factor each trinomial.**

1) $x^2 + 8x + 12 =$

2) $x^2 - 6x + 5 =$

3) $x^2 + 15x + 36 =$

4) $x^2 - 12x + 35 =$

5) $x^2 - 11x + 18 =$

6) $x^2 - 9x + 18 =$

7) $x^2 + 18x + 72 =$

8) $x^2 - x - 72 =$

9) $x^2 + 4x - 21 =$

10) $x^2 - 13x + 22 =$

11) $x^2 + 2x - 24 =$

12) $x^2 - 3x - 40 =$

13) $x^2 - 3x - 70 =$

14) $x^2 + 26x + 169 =$

15) $4x^2 - 7x - 15 =$

16) $x^2 - 14x + 33 =$

17) $10x^2 + 5x - 15 =$

18) $6x^2 - 4x - 42 =$

19) $x^2 + 12x + 36 =$

20) $5x^2 + 17x - 12 =$

✎ **Calculate each problem.**

21) The area of a rectangle is $x^2 - x - 56$. If the width of rectangle is $x + 7$, what is its length? _____

22) The area of a parallelogram is $4x^2 + 17x - 15$ and its height is $x + 5$. What is the base of the parallelogram? _____

23) The area of a rectangle is $6x^2 - 22x + 12$. If the width of the rectangle is $3x - 2$, what is its length? _____

WWW.MathNotion.Com

TASC Subject Test – Mathematics

Operations with Polynomials

✍ **Find each product.**

1) $4(5x + 3) =$ _____

2) $8(2x + 6) =$ _____

3) $2(5x - 2) =$ _____

4) $-4(7x - 3) =$ _____

5) $3x^2(9x + 1) =$ _____

6) $4x^6(7x - 9) =$ _____

7) $3x^4(-7x + 3) =$ _____

8) $-8x^4(5x - 8) =$ _____

9) $7(x^2 + 5x - 3) =$ _____

10) $9(5x^2 - 7x + 5) =$ _____

11) $3(3x^2 + 3x + 2) =$ _____

12) $5x(3x^2 + 5x + 8) =$ _____

13) $(5x + 7)(3x - 3) =$ _____

14) $(9x + 3)(3x - 5) =$ _____

15) $(6x + 3)(4x - 2) =$ _____

16) $(7x - 2)(3x + 5) =$ _____

✍ **Calculate each problem.**

17) The measures of two sides of a triangle are $(2x + 5y)$ and $(6x - 3y)$. If the perimeter of the triangle is $(13x + 4y)$, what is the measure of the third side? _____

18) The height of a triangle is $(8x + 5)$ and its base is $(4x - 3)$. What is the area of the triangle? _____

19) One side of a square is $(6x + 2)$. What is the area of the square? _____

20) The length of a rectangle is $(5x - 8y)$ and its width is $(15x + 8y)$. What is the perimeter of the rectangle? _____

21) The side of a cube measures $(x + 2)$. What is the volume of the cube? _____

22) If the perimeter of a rectangle is $(28x + 6y)$ and its width is $(5x + 2y)$, what is the length of the rectangle? _____

WWW.MathNotion.Com

TASC Subject Test – Mathematics

Answers of Worksheets

Writing Polynomials in Standard Form

1) $4x$
2) -5
3) $6x^5 - 12x^3$
4) $14x^4$
5) $-9x^3 + 5x^2 + 4x$
6) $12x^5 - 3x^2$
7) $-2x^8 + 8x^3 + 5x$
8) $-9x^6 - 7x^3 + 4x$
9) $3x^2 - 6x + 22$
10) $9x^4 - 4x + 3$
11) $-8x^3 + 13x^2 + 28x$
12) $-2x^3 + 4x^2 + 16$
13) $9x^4 + 19x^2 - 9x$
14) $3x^4 - 2x^3 - 7x^2$
15) $-8x^4 + 3x^2 - 51$
16) $-8x^6 + 4x^4 + 7x^2 - 15$
17) $-4x^5 + 6x^4 - 3x^3 + 16$
18) $-2x^6 - 7x^2 - x$
19) $6x^7 + 8x^5 - 3x^2$
20) $3x^6 - 12x^5 + 10x^2$
21) $-11x^7 + 20x^5 - 6x^3$
22) $-5x^5 + 34x^4 + 6x^3$
23) $-6x^3 + 12x^2 + 15x$
24) $12x^7 + 48x^4$
25) $15x^3 + 30x^2 + 20x$
26) $42x^6 - 14x^2 + 28x$
27) $12x^5 - 12x^4 + 6x$
28) $8x^6 + 24x^3 - 12x$
29) $15x^5 + 20x^4 + 10x^2$
30) $8x^7 - 4x^4 + 6x^2$

Simplifying Polynomials

1) $12x - 60$
2) $15x^2 - 20x$
3) $30x^2 - 42x$
4) $21x^2 + 15x$
5) $20x^2 - 15x$
6) $48x^2 + 12x$
7) $3x^2 - 14x + 8$
8) $2x^2 - 4x - 30$
9) $x^2 - 10x + 21$
10) $9x^2 - 16$
11) $25x^2 - 30x + 8$
12) $-8x^4 + 12x^2$
13) $5x^3 - 2x^2 + 3x + 7$
14) $-10x^3 + 4x^2 + 7x$
15) $5x^5 - 6x^3 + 12x^2$
16) $6x^8 + 4x^6 - 5x^2$
17) $-4x^6 + 10x^5 - 12x^3 + 4x$
18) $-16x^3 - 3x^2 + 22$
19) $4x^3 + 2x^2 - 4x$
20) $-x^5 - 4x^2 + 18$
21) $x^6 - 14x^3$
22) $12x^4 - 4$
23) $-9x^4 - 12x^3$
24) -90

TASC Subject Test – Mathematics

25) $-9x^4 - 2x + 19$

26) $-7x^5 - 2x^3 + 17$

27) $11x^3 - 10x + 2$

28) $-8x^4 + 2x^2$

29) $2x^5 + 6x + 2$

30) $11x^5 - 5x^4 + 4$

31) $-9x^5 - 28$

32) $9x^3 - 6x$

Adding and Subtracting Polynomials

1) $x^2 + 1$

2) $6x^3 - 1$

3) $2x^5 + 5x^2 - 15$

4) $6x^3 + 3x^2 - 4x$

5) $-24x^4 + 28x - 6$

6) $14x^2$

7) $14x^2 - 6$

8) $6x^2 - 4$

9) $-3x^3 + 2x$

10) $3x$

11) $19x^3 - 6x$

12) $-7x^5 - 4x$

13) $-11x^7 + 2x^2 + 6x$

14) $4x^5 + 11x^2 - 5$

15) $4x^5 + 18x^4$

16) $-9x^3 + 7x$

17) $-151x^2 + 3x$

18) $-11x^4 + 4x^2 + 5x$

19) $10x^4 - 3x^2 - 4$

20) $10x^6 + 6x^3$

21) $28x^5 + 17x^4 + 22x^2$

22) $10x^4 - 12x$

23) $32x^4 + 4x$

24) $5x^8 - 15x^6 - 5x^3 + 3x$

25) $6x^4 + 15x^2 - 12x$

26) $-2x^9 + 13x^7 + 40x^3$

Multiplying Monomials

1) $-6u^{10}$

2) $10p^{11}$

3) $12xy^3z^9$

4) $24u^6t^5$

5) $35a^5b^6$

6) $-18a^6b^4$

7) $13x^5y^9$

8) $-48p^5q^9$

9) $32s^5t^6$

10) $-36x^6y^4$

11) $36xy^7z^4$

12) $24px^3y^2$

13) $-39p^3q^5$

14) $13s^4t^8$

15) $-66p^8$

16) $-24p^4q^9r^7$

17) $28a^7b$

18) $-30u^9v^6$

19) $-27u^6$

20) $-24x^3y^6$

21) $-13y^8z^4$

22) $16a^5b^2c^6$

23) $35p^9q^8$

24) $-16u^{12}v^6$

25) $-17y^{10}z^6$

26) $-40p^3q^7r^3$

27) $15a^5b^6c^8$

28) $18x^5y^8z^5$

Multiplying and Dividing Monomials

1) $10x^7$

2) $24x^6$

3) $21x^8$

4) $20x^8$

5) $36x^{10}$

6) $32x^{11}y^5$

7) $14x^7y^6$

8) $-10x^6y^9$

9) $18x^7y^7$

10) $-25x^5y^4$

11) $24x^7y^7$

12) $20x^5y^6$

WWW.MathNotion.Com

TASC Subject Test – Mathematics

13) $48x^7y^{16}$
14) $45x^7y^{11}$
15) $56x^8y^{14}$
16) $-21x^{12}y^{12}$
17) $5x^5y^2$

18) xy^4
19) $7x^3y^3$
20) $9xy$
21) $4x^5y^3$
22) $12x^5y$

23) $8x^{10}y^3$
24) $5x^{-1}y^3$
25) $5x^3$
26) $-7x^{14}y^5$
27) $-4x^2$

Multiplying a Polynomial and a Monomial

1) $2x^2 + 4x$
2) $-12x + 24$
3) $20x^2 + 10x$
4) $-4x^2 + 5x$
5) $16x^2 - 16x$
6) $12x - 24y$
7) $35x^2 - 35x$
8) $36x^2 + 6xy$
9) $4x^2 + 24xy$
10) $33x^2 + 44xy$
11) $21x^2 + 14x$
12) $40x^2 - 100xy$
13) $27x^2 - 18xy$
14) $7x^2 - 28xy + 42x$
15) $16x^3 + 40xy^2$
16) $24x^2 + 36xy$

17) $8x^4 - 16y^4$
18) $-12x^3y + 16xy$
19) $-20x^3 + 8xy - 16$
20) $4x^2 - 20xy - 24$
21) $16x^4 - 40x^2y + 16x^2$
22) $12x^4 + 36x^2 - 12x^2y$
23) $6x^2 + 3xy - 27y^2$
24) $20x^4 - 12x^2 + 28x$
25) $18x^{22} - 12x - 30$
26) $-2x^5 + 4x^3 + 3x^2$
27) $4x^5 - 2x^3 + 10x^2$
28) $12x^7 - 8x^5 + 20x^4$
29) $8x^6 - 10x^3y + 14x^2y^3$
30) $25x^6 - 15x^3 + 45x^2$
31) $42x^4 + 21x^3 - 42x^2$
32) $4x^4 - 16x^2y + 8xy^2$

Multiplying Binomials

1) $x^2 + 9x + 18$
2) $x^2 - x - 12$
3) $x^2 - 11x + 24$
4) $x^2 + 17x + 72$
5) $x^2 - 14x + 24$
6) $x^2 + 10x + 25$
7) $x^2 + x - 42$

8) $x^2 - 11x + 24$
9) $x^2 + 19x + 84$
10) $x^2 + 4x - 32$
11) $x^2 + 16x + 64$
12) $x^2 + 9x + 14$
13) $x^2 - 36$
14) $x^2 - 25$

TASC Subject Test – Mathematics

15) $x^2 + 22x + 121$
16) $x^2 + 15x + 54$
17) $x^2 - 4$
18) $x^2 + 3x - 28$
19) $3x^2 + 23x + 30$
20) $20x^2 + 16x - 48$
21) $3x^2 - 14x - 49$
22) $x^2 - 13x + 36$
23) $x^2 - 10x - 24$

24) $10x^2 - 12x - 16$
25) $3x^2 + 16x - 64$
26) $42x^2 + 9x - 6$
27) $12x^2 + 35x + 25$
28) $63x^2 - 8x - 16$
29) $2x^2 - 4x - 16$
30) $25x^2 - 16$
31) $9x^2 - 15x - 14$
32) $x^4 - 64$

Factoring Trinomials

1) $(x+6)(x+2)$
2) $(x-5)(x-1)$
3) $(x+12)(x+3)$
4) $(x-5)(x-7)$
5) $(x-2)(x-9)$
6) $(x-6)(x-3)$
7) $(x+6)(x+12)$
8) $(x+8)(x-9)$

9) $(x-3)(x+7)$
10) $(x-11)(x-2)$
11) $(x-4)(x+6)$
12) $(x-8)(x+5)$
13) $(x+7)(x-10)$
14) $(x+13)(x+13)$
15) $(4x+5)(x-3)$
16) $(x-11)(x-3)$

17) $(5x-5)(2x+3)$
18) $(2x-6)(3x+7)$
19) $(x+6)(x+6)$
20) $(5x-3)(x+4)$
21) $(x-8)$
22) $(4x-3)$
23) $(2x-6)$

Operations with Polynomials

1) $20x + 12$
2) $16x + 48$
3) $10x - 4$
4) $-28x + 12$
5) $27x^3 + 3x^2$
6) $28x^7 - 36x^6$
7) $-21x^5 + 9x^4$
8) $-40x^5 + 64x^4$

9) $7x^2 + 35x - 21$
10) $45x^2 - 63x + 45$
11) $9x^2 + 9x + 6$
12) $15x^3 + 25x^2 + 40x$
13) $15x^2 + 6x - 21$
14) $27x^2 - 36x - 15$
15) $24x^2 - 6$
16) $21x^2 + 29x - 10$

17) $(5x + 2y)$
18) $16x^2 - 2x - \frac{15}{2}$
19) $36x^2 + 24x + 4$
20) $40x$
21) $x^3 + 6x^2 + 12x + 8$
22) $(9x + y)$

TASC Subject Test – Mathematics

Chapter 9 : Functions Operations and Quadratic

Topics that you'll practice in this chapter:

- ✓ Evaluating Function
- ✓ Adding and Subtracting Functions
- ✓ Multiplying and Dividing Functions
- ✓ Composition of Functions
- ✓ Quadratic Equation
- ✓ Solving Quadratic Equations
- ✓ Quadratic Formula and the Discriminant
- ✓ Graphing Quadratic Functions

It's fine to work on any problem, so long as it generates interesting mathematics along the way – even if you don't solve it at the end of the day." – Andrew Wiles

TASC Subject Test – Mathematics

Evaluating Function

✎ **Write each of following in function notation.**

1) $h = -8x + 3$

2) $k = 2a - 14$

3) $d = 11t$

4) $y = \frac{5}{12}x - \frac{7}{12}$

5) $m = 24n - 210$

6) $c = p^2 - 5p + 10$

✎ **Evaluate each function.**

7) $f(x) = 2x - 7$, find $f(-3)$

8) $g(x) = \frac{1}{9}x + 12$, find $f(18)$

9) $h(x) = -4x + 9$, find $f(3)$

10) $f(x) = -x + 19$, find $f(-3)$

11) $f(a) = 7a - 12$, find $f(3)$

12) $h(x) = 14 - 3x$, find $f(-4)$

13) $g(n) = 6n - 10$, find $f(2)$

14) $f(x) = -11x - 4$, find $f(-1)$

15) $k(n) = -20 - 3.5n$, find $f(2)$

16) $f(x) = -0.7x + 3.3$, find $f(-7)$

17) $g(n) = \frac{11n+8}{n}$, find $g(2)$

18) $g(n) = \sqrt{3n} + 12$, find $g(3)$

19) $h(x) = x^{-2} - 7$, find $h(\frac{1}{9})$

20) $h(n) = n^{-3} + 11$, find $h(\frac{1}{4})$

21) $h(n) = n^3 - 2$, find $h(\frac{1}{2})$

22) $h(n) = n^2 - 4$, find $h(-\frac{1}{3})$

23) $h(n) = 4n^2 - 13$, find $h(-5)$

24) $h(n) = -2n^3 - 6n$, find $h(2)$

25) $g(n) = \sqrt{16n^2} - \sqrt{n}$, find $g(4)$

26) $h(a) = \frac{-14a+9}{3a}$, find $h(-b)$

27) $k(a) = 12a - 14$, find $k(a-3)$

28) $h(x) = \frac{1}{9}x + 18$, find $h(-18x)$

29) $h(x) = 8x^2 + 16$, find $h(\frac{x}{2})$

30) $h(x) = x^4 - 20$, find $h(-2x)$

TASC Subject Test – Mathematics

Adding and Subtracting Functions

✎ **Perform the indicated operation.**

1) $f(x) = 2x + 3$

   $g(x) = x + 7$

   Find $(f - g)(2)$

2) $g(a) = -5a - 8$

   $f(a) = -3a - 5$

   Find $(g - f)(-2)$

3) $h(t) = 4t + 3$

   $g(t) = 4t + 7$

   Find $(h - g)(t)$

4) $g(a) = -6a - 10$

   $f(a) = 3a^2 + 9$

   Find $(g - f)(x)$

5) $g(x) = \frac{5}{6}x - 23$

   $h(x) = \frac{5}{12}x + 25$

   Find $g(12) - h(12)$

6) $h(x) = \sqrt{3x} - 2$

   $g(x) = \sqrt{3x} + 5$

   Find $(h + g)(12)$

7) $f(x) = x^{-1}$

   $g(x) = x^2 + \frac{5}{x}$

   Find $(f - g)(-3)$

8) $h(n) = n^2 + 2$

   $g(n) = -4n + 6$

   Find $(h - g)(2a)$

9) $g(x) = -2x^2 - 5 - 4x$

   $f(x) = 7 + 2x$

   Find $(g - f)(3x)$

10) $g(t) = 11t - 4$

    $f(t) = -2t^2 + 5$

    Find $(g + f)(-t)$

11) $f(x) = 8x + 9$

    $g(x) = -5x^2 + 3x$

    Find $(f - g)(-x^2)$

12) $f(x) = -3x^4 - 5x$

    $g(x) = 2x^4 + 5x + 22$

    Find $(f + g)(3x^2)$

WWW.MathNotion.Com

> TASC Subject Test – Mathematics

Multiplying and Dividing Functions

✎ **Perform the indicated operation.**

1) $g(x) = -2x - 1$
   $f(x) = 4x + 3$
   Find $(g \cdot f)(2)$

2) $f(x) = 5x$
   $h(x) = -2x + 3$
   Find $(f \cdot h)(-2)$

3) $g(a) = 5a - 2$
   $h(a) = 2a - 3$
   Find $(g \cdot h)(-3)$

4) $f(x) = 2x - 7$
   $h(x) = x - 5$
   Find $\left(\dfrac{f}{h}\right)(4)$

5) $f(x) = 8a^2$
   $g(x) = 3 + 2a$
   Find $\left(\dfrac{f}{g}\right)(2)$

6) $g(a) = \sqrt{4a} + 2$
   $f(a) = (-a)^4 + 1$
   Find $\left(\dfrac{g}{f}\right)(1)$

7) $g(t) = t^3 + 1$
   $h(t) = 5t - 2$
   Find $(g \cdot h)(-2)$

8) $g(n) = n^2 + 2n - 4$
   $h(n) = -5n + 3$
   Find $(g \cdot h)(1)$

9) $g(a) = (a - 3)^2$
   $f(a) = a^2 + 4$
   Find $\left(\dfrac{g}{f}\right)(3)$

10) $g(x) = -3x^2 + \dfrac{4}{5}x + 9$
    $f(x) = x^2 - 24$
    Find $\left(\dfrac{g}{f}\right)(5)$

11) $f(x) = 2x^3 - 5x^2 + 1$
    $g(x) = 3x - 1$
    Find $(f \cdot g)(x)$

12) $f(x) = 5x - 2$
    $g(x) = x^3 - 2x$
    Find $(f \cdot g)(x^2)$

WWW.MathNotion.Com

TASC Subject Test – Mathematics

Composition of Functions

Using $f(x) = 2x - 5$ and $g(x) = -2x$, find:

1) $f(g(2)) =$

2) $f(g(-1)) =$

3) $g(f(-4)) =$

4) $g(f(5)) =$

5) $f(g(3)) =$

6) $g(f(0)) =$

Using $f(x) = -\frac{1}{4}x + \frac{3}{4}$ and $g(x) = 2x^2$, find:

7) $g(f(-2)) =$

8) $g(f(4)) =$

9) $g(g(1)) =$

10) $f(f(1)) =$

11) $g(f(-4)) =$

12) $f(g(x)) =$

Using $f(x) = -2x + 2$ and $g(x) = x + 1$, find:

13) $g(f(1)) =$

14) $f(f(0)) =$

15) $f(g(-1)) =$

16) $f(g(-3)) =$

17) $g(f(2)) =$

18) $f(g(x)) =$

Using $f(x) = \sqrt{x + 9}$ and $g(x) = x - 9$, find:

19) $f(g(9)) =$

20) $g(f(-9)) =$

21) $f(g(4)) =$

22) $f(f(7)) =$

23) $g(f(-5)) =$

24) $g(g(0)) =$

TASC Subject Test – Mathematics

Quadratic Equation

✎ Multiply.

1) $(x - 4)(x + 6) =$ _____

2) $(x + 5)(x + 7) =$ _____

3) $(x - 6)(x + 8) =$ _____

4) $(x + 2)(x - 9) =$ _____

5) $(x - 7)(x - 8) =$ _____

6) $(3x + 2)(x - 3) =$ _____

7) $(4x - 3)(x + 2) =$ _____

8) $(4x - 5)(x + 1) =$ _____

9) $(7x + 1)(x - 6) =$ _____

10) $(5x + 1)(3x - 3) =$ _____

✎ Factor each expression.

11) $x^2 - 2x - 8 =$ _____

12) $x^2 + 8x + 15 =$ _____

13) $x^2 - 2x - 24 =$ _____

14) $x^2 - 10x + 21 =$ _____

15) $x^2 + 10x + 21 =$ _____

16) $4x^2 + 9x + 5 =$ _____

17) $5x^2 + 13x - 6 =$ _____

18) $5x^2 + 17x - 12 =$ _____

19) $2x^2 + 7x + 5 =$ _____

20) $9x^2 - 21x + 6 =$ _____

✎ Calculate each equation.

21) $(x + 6)(x - 3) = 0$

22) $(x + 1)(x + 8) = 0$

23) $(3x + 6)(x + 5) = 0$

24) $(2x - 2)(4x + 8) = 0$

25) $x^2 + x + 10 = 22$

26) $x^2 + 11x + 36 = 12$

27) $2x^2 + 9x + 9 = 5$

28) $x^2 + 3x - 24 = 4$

29) $5x^2 + 5x - 40 = 20$

30) $8x^2 + 8x = 48$

TASC Subject Test – Mathematics

Solving Quadratic Equations

✏ **Solve each equation by factoring or using the quadratic formula.**

1) $(x+9)(x-1) = 0$

2) $(x+7)(x+6) = 0$

3) $(x-8)(x+3) = 0$

4) $(x-6)(x-4) = 0$

5) $(x+2)(x+12) = 0$

6) $(5x+4)(x+7) = 0$

7) $(6x+1)(4x+5) = 0$

8) $(2x+7)(x+8) = 0$

9) $(x+6)(3x+15) = 0$

10) $(12x+2)(x+8) = 0$

11) $x^2 = 8x$

12) $x^2 - 16 = 0$

13) $3x^2 + 6 = 9x$

14) $-2x^2 - 8 = 10x$

15) $5x^2 + 40x = 45$

16) $x^2 + 10x = 24$

17) $x^2 + 6x = 16$

18) $x^2 + 9x = -18$

19) $x^2 + 13x = -36$

20) $x^2 + 3x - 15 = 5x$

21) $x^2 + 8x + 7 = -8$

22) $3x^2 - 11x = -9 + x$

23) $10x^2 + 3 = 27x - 15$

24) $7x^2 - 6x + 8 = 8$

25) $2x^2 - 12 = -3x + 2$

26) $10x^2 - 26x - 3 = -15$

27) $3x^2 + 21 = -16x + 5$

28) $x^2 + 15x - 10 = -66$

29) $3x^2 - 8x - 8 = 4 + x$

30) $2x^2 + 6x - 24 = 12$

31) $3x^2 - 33x + 54 = -18$

32) $-10x^2 - 15x - 9 = -9 - 27x^2$

WWW.MathNotion.Com

TASC Subject Test – Mathematics

Quadratic Formula and the Discriminant

✍ Find the value of the discriminant of each quadratic equation.

1) $3x(x - 8) = 0$

2) $2x^2 + 6x - 4 = 0$

3) $x^2 + 6x + 7 = 0$

4) $x^2 - x + 3 = 0$

5) $x^2 + 4x - 3 = 0$

6) $2x^2 + 6x - 10 = 0$

7) $3x^2 + 7x + 5 = 0$

8) $x^2 - 6x - 4 = 0$

9) $2x^2 + 8x + 3 = 0$

10) $x^2 + 7x - 5 = 0$

11) $5x^2 + 2x - 3 = 0$

12) $-3x^2 - 11x + 4 = 0$

13) $-6x^2 - 12x + 8 = 0$

14) $-x^2 - 9x - 12 = 0$

15) $7x^2 - 6x - 10 = 0$

16) $-4x^2 - 2x + 8 = 0$

17) $5x^2 + 8x - 2 = 0$

18) $6x^2 - 4x = 0$

19) $3x^2 - 5x + 2 = 0$

20) $4x^2 + 9x + 3 = 0$

✍ Find the discriminant of each quadratic equation then state the number of real and imaginary solutions.

21) $-4x^2 - 16 = 16x$

22) $20x^2 = 20x - 5$

23) $-11x^2 - 19x = 26$

24) $22x^2 - 4x + 1 = 18x^2$

25) $-11x^2 = -15x + 8$

26) $3x^2 + 6x + 9 = 6$

27) $13x^2 - 5x - 12 = -26$

28) $-8x^2 - 32x - 25 = 7$

WWW.MathNotion.Com

TASC Subject Test – Mathematics

Graphing Quadratic Functions

✎ketch the graph of each function. Identify the vertex and axis of symmetry.

1) $y = (x + 3)^2 + 2$

2) $y = (x - 3)^2 - 2$

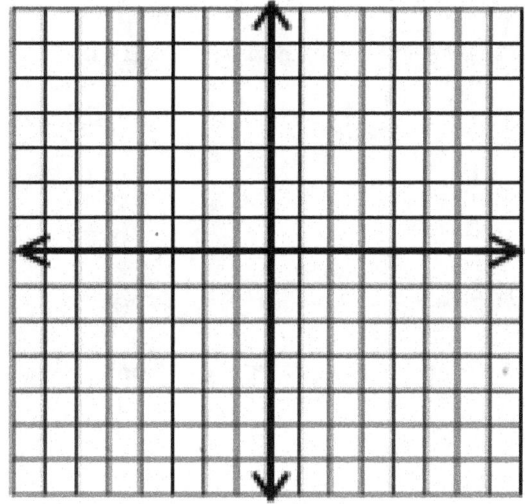

3) $y = 6 - (-x + 4)^2$

4) $y = -3x^2 - 6x + 9$

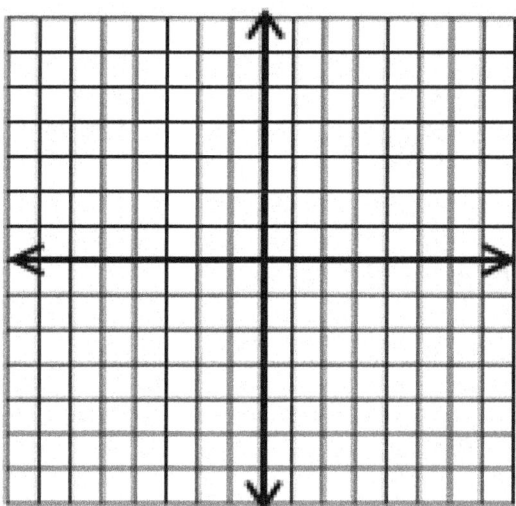

TASC Subject Test – Mathematics

Answers of Worksheets

Evaluating Function

1) $h(x) = -8x + 3$
2) $k(a) = 2a - 14$
3) $d(t) = 11t$
4) $f(x) = \frac{5}{12}x - \frac{7}{12}$
5) $m(n) = 24n - 210$
6) $c(p) = p^2 - 5p + 10$
7) -13
8) 14
9) -3
10) 22
11) 9
12) 26
13) 2
14) 7
15) -27
16) 8.2
17) 15
18) 15
19) 74
20) 75
21) $-1\frac{7}{8}$
22) $-3\frac{8}{9}$
23) 87
24) -28
25) 14
26) $-\frac{14b+9}{3b}$
27) $12a - 50$
28) $-2x + 18$
29) $2x^2 + 16$
30) $16x^4 - 20$

Adding and Subtracting Functions

1) -2
2) 1
3) -4
4) $-3x^2 - 6x - 19$
5) -43
6) 15
7) $-7\frac{2}{3}$
8) $4a^2 + 8a - 4$
9) $-18x^2 - 18x - 12$
10) $-2t^2 - 11t + 1$
11) $5x^4 - 5x^2 + 9$
12) $-81x^8 + 22$

Multiplying and Dividing Functions

1) -55
2) -70
3) 153
4) -1
5) $4\frac{4}{7}$
6) 2
7) 84
8) 2
9) 0
10) -62
11) $6x^4 - 17x^3 + 5x^2 + 3x - 1$
12) $5x^8 - 2x^6 - 10x^4 + 4x^2$

Composition of Functions

1) -13
2) -1
3) 26
4) -10
5) -17
6) 10
7) $\frac{25}{8}$
8) $\frac{1}{8}$
9) 8
10) $\frac{5}{8}$
11) $\frac{49}{8}$
12) $-\frac{1}{2}(x^2 - \frac{3}{2})$
13) 1
14) -2
15) 2
16) 6
17) -1
18) $-2x$

WWW.MathNotion.Com

TASC Subject Test – Mathematics

19) 3
20) −9
21) 2
22) $\sqrt{13}$
23) −7
24) −18

Quadratic Equations

1) $x^2 + 2x - 24$
2) $x^2 + 12x + 35$
3) $x^2 + 2x - 48$
4) $x^2 - 7x - 18$
5) $x^2 - 15x + 56$
6) $3x^2 - 7x - 6$
7) $4x^2 + 5x - 6$
8) $4x^2 - x - 5$
9) $7x^2 - 41x - 6$
10) $15x^2 - 12x - 3$
11) $(x - 4)(x + 2)$
12) $(x + 5)(x + 3)$
13) $(x - 6)(x + 4)$
14) $(x - 3)(x - 7)$
15) $(x + 3)(x + 7)$
16) $(4x + 5)(x + 1)$
17) $(5x - 2)(x + 3)$
18) $(5x - 3)(x + 4)$
19) $(2x + 5)(x + 1)$
20) $3(x - 2)(3x - 1)$
21) $x = -6, x = 3$
22) $x = -1, x = -8$
23) $x = -2, x = -5$
24) $x = 1, x = -2$
25) $x = 3, x = -4$
26) $x = -3, x = -8$
27) $x = -4, x = -\frac{1}{2}$
28) $x = 4, x = -7$
29) $x = 3, x = -4$
30) $x = -3, x = 2$

Solving quadratic equations

1) $\{-9, 1\}$
2) $\{-6, -7\}$
3) $\{8, -3\}$
4) $\{6, 4\}$
5) $\{-2, -12\}$
6) $\{-\frac{4}{5}, -7\}$
7) $\{-\frac{5}{4}, -\frac{1}{6}\}$
8) $\{-\frac{7}{2}, -8\}$
9) $\{-6, -5\}$
10) $\{-\frac{1}{6}, -8\}$
11) $\{8, 0\}$
12) $\{4, -4\}$
13) $\{2, 1\}$
14) $\{-4, -1\}$
15) $\{1, -9\}$
16) $\{2, -12\}$
17) $\{2, -8\}$
18) $\{-3, -6\}$
19) $\{-4, -9\}$
20) $\{5, -3\}$
21) $\{-5, -3\}$
22) $\{1, 3\}$
23) $\{\frac{6}{5}, \frac{3}{2}\}$
24) $\{\frac{6}{7}, 0\}$
25) $\{-\frac{7}{2}, 2\}$
26) $\{\frac{3}{5}, 2\}$
27) $\{-\frac{4}{3}, -4\}$
28) $\{-8, -7\}$
29) $\{4, -1\}$
30) $\{3, -6\}$
31) $\{3, 8\}$
32) $\{\frac{15}{17}, 0\}$

Quadratic formula and the discriminant

1) 576
2) 68
3) 8
4) −11
5) 28
6) 116
7) −11
8) 52
9) 40
10) 69
11) 64
12) 169
13) 336
14) 33
15) 316
16) 132
17) 104
18) 16
19) 1
20) 33
21) 0, one real solution
22) 0, one real solution
23) −783, no solution

TASC Subject Test – Mathematics

24) 0, *one real solution* 26) 0, *one real solution* 28) 0, *one real solution*

25) −127, *no solution* 27) −703, *no solution*

Graphing quadratic functions

1) $(-3, 2), x = -3$

2) $(3, -2), x = 3$

3) $(4, 6), x = 4$

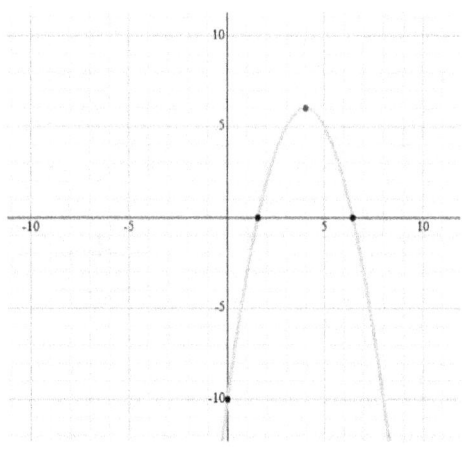

4) $(-1, 12), x = -1$

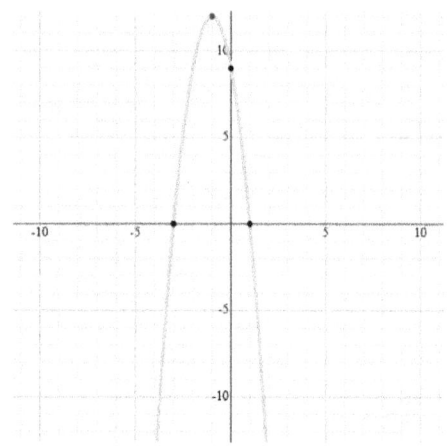

TASC Subject Test – Mathematics

Chapter 10 : Geometry and Solid Figures

Topics that you'll practice in this chapter:

- ✓ Angles
- ✓ Pythagorean Relationship
- ✓ Triangles
- ✓ Polygons
- ✓ Trapezoids
- ✓ Circles
- ✓ Cubes
- ✓ Rectangular Prism
- ✓ Cylinder
- ✓ Pyramids and Cone

Mathematics is, as it were, a sensuous logic, and relates to philosophy as do the arts, music, and plastic art to poetry. — K. Shegel

TASC Subject Test – Mathematics

Angles

✎ **What is the value of *x* in the following figures?**

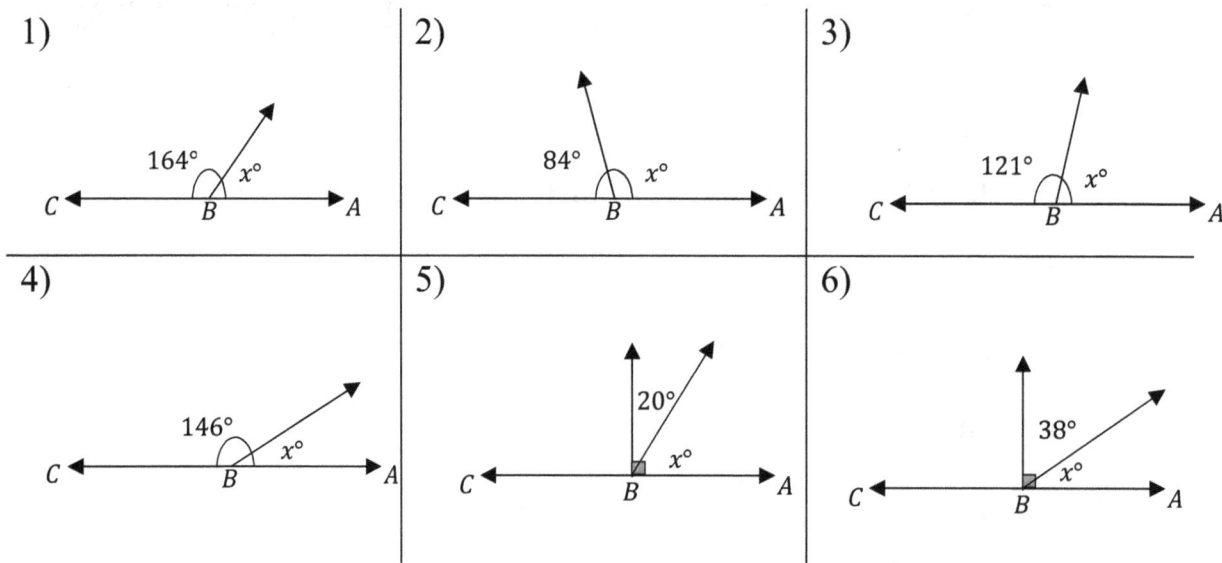

✎ **Calculate.**

7) Two supplement angles have equal measures. What is the measure of each angle? _____

8) The measure of an angle is seven fifth the measure of its supplement. What is the measure of the angle? _____

9) Two angles are complementary and the measure of one angle is 24 less than the other. What is the measure of the smaller angle? _____

10) Two angles are complementary. The measure of one angle is one fifth the measure of the other. What is the measure of the bigger angle? _____

11) Two supplementary angles are given. The measure of one angle is 40° less than the measure of the other. What does the smaller angle measure? _____

WWW.MathNotion.Com

TASC Subject Test – Mathematics

Pythagorean Relationship

✎ Do the following lengths form a right triangle?

1)

2)

3)

4)

5)

6)

7)

8)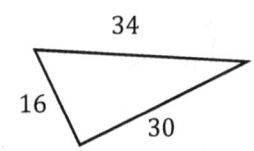

✎ Find the missing side?

9)

10)

11)

12)

13)

14)

15)

16)

WWW.MathNotion.Com   125

TASC Subject Test – Mathematics

Triangles

✏️ Find the measure of the unknown angle in each triangle.

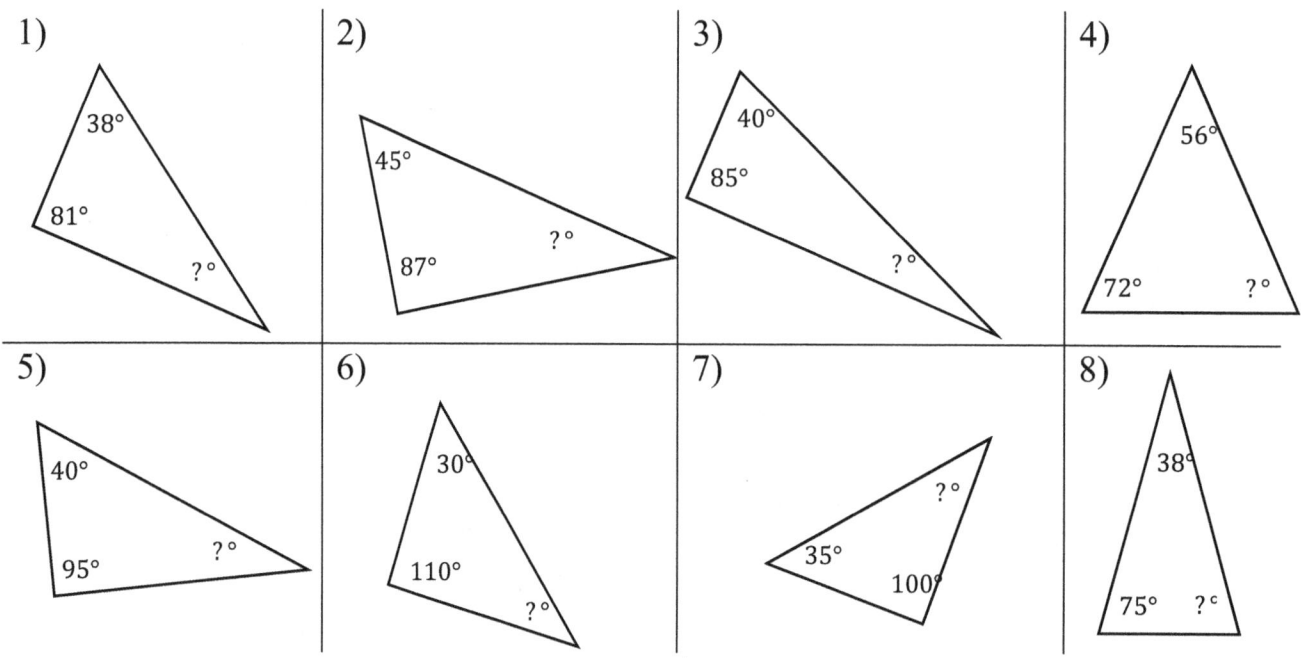

✏️ Find area of each triangle.

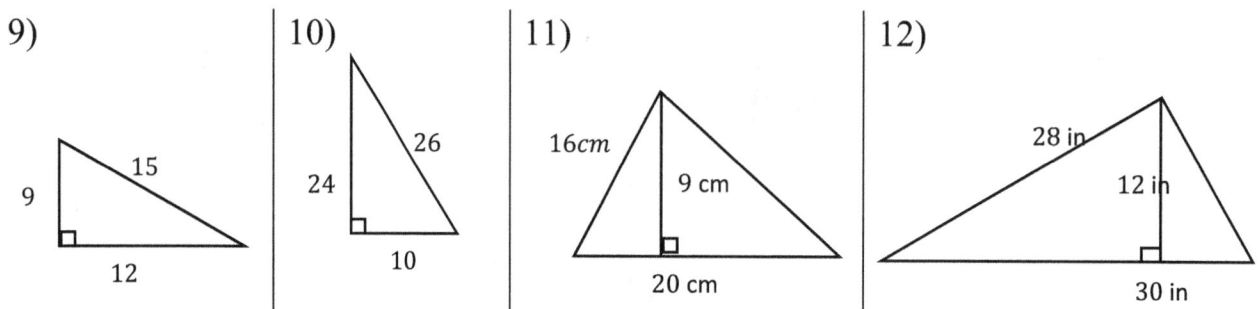

TASC Subject Test – Mathematics

Polygons

✏ **Find the perimeter of each shape.**

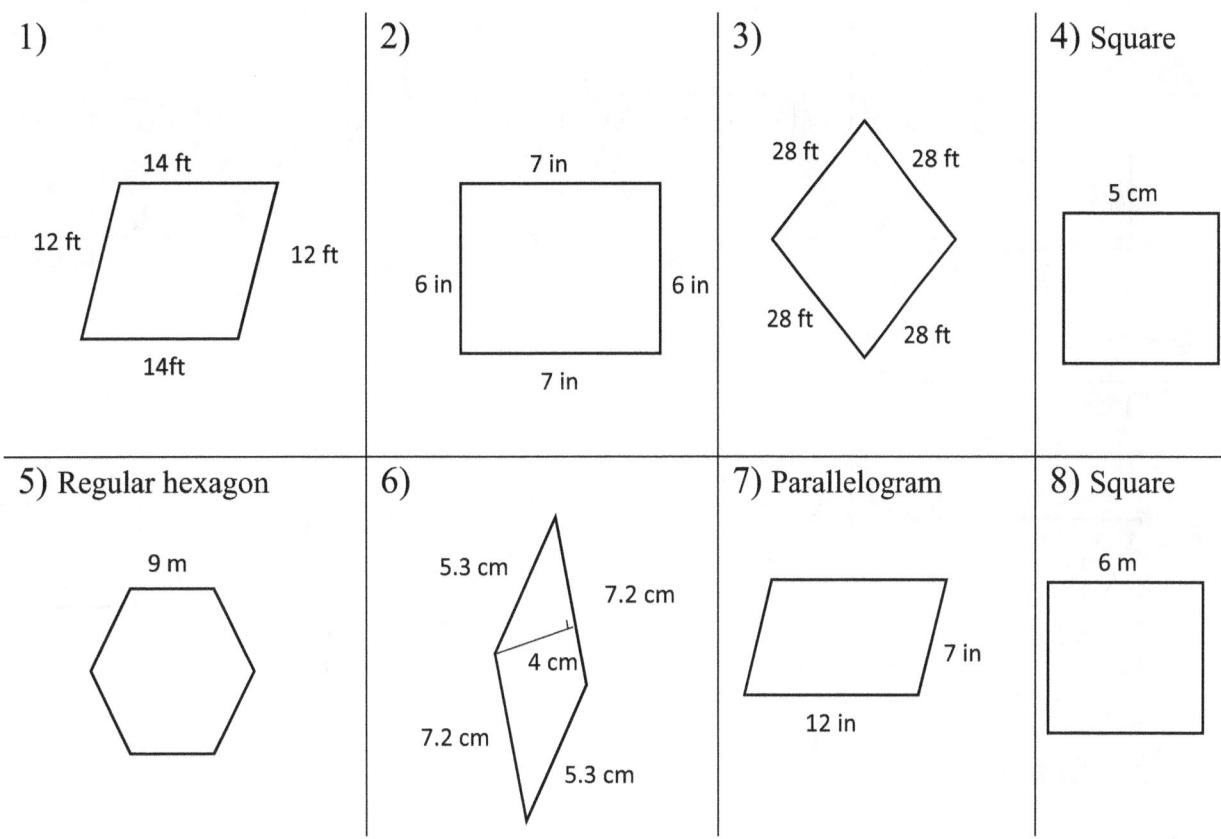

✏ **Find the area of each shape.**

9) Parallelogram — 5 m, 6 m, 5 m

10) Rectangle — 25 in, 12 in

11) Rectangle — 16 km, 10 km

12) Square — 7 in

TASC Subject Test – Mathematics

Trapezoids

✎ Find the area of each trapezoid.

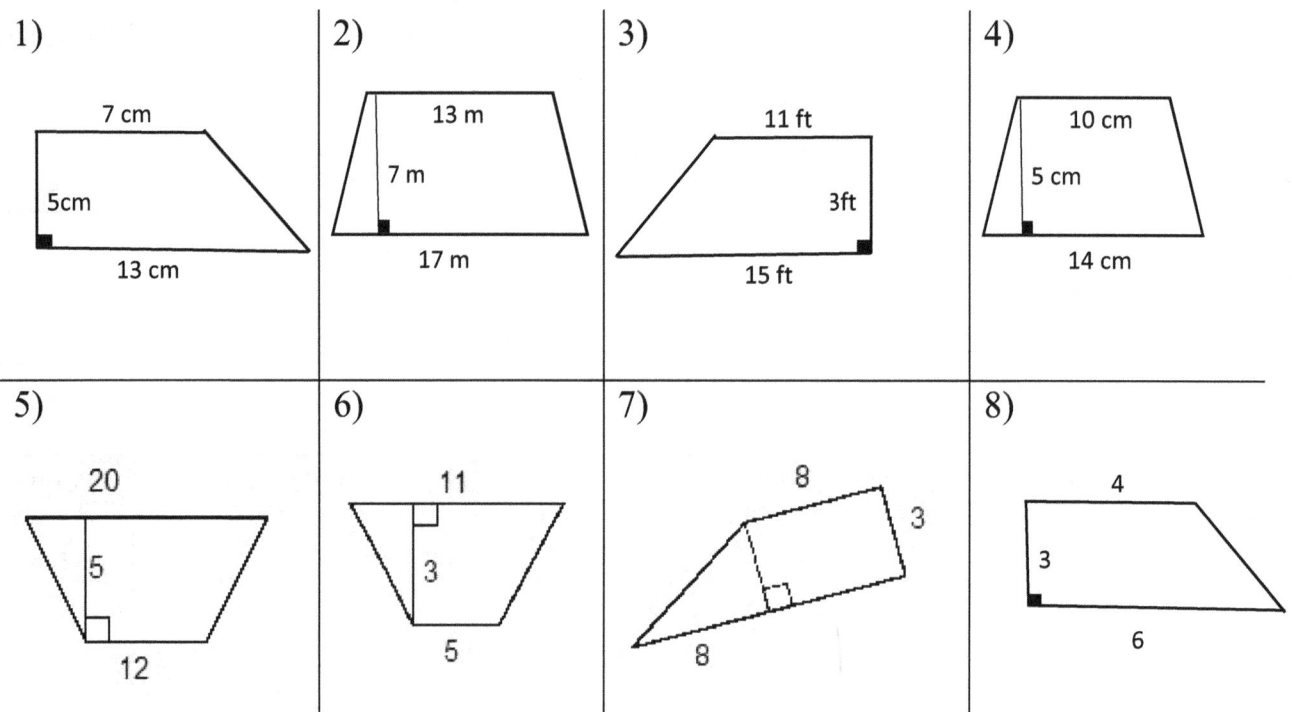

✎ Calculate.

1) A trapezoid has an area of 45 cm² and its height is 5 cm and one base is 5 cm. What is the other base length? _____

2) If a trapezoid has an area of 99 ft² and the lengths of the bases are 8 ft and 10 ft, find the height? _____

3) If a trapezoid has an area of 126 m² and its height is 14 m and one base is 6 m, find the other base length? _____

4) The area of a trapezoid is 440 ft² and its height is 22 ft. If one base of the trapezoid is 15 ft, what is the other base length? _____

WWW.MathNotion.Com

Circles

✎ **Find the area of each circle.** (π = 3.14)

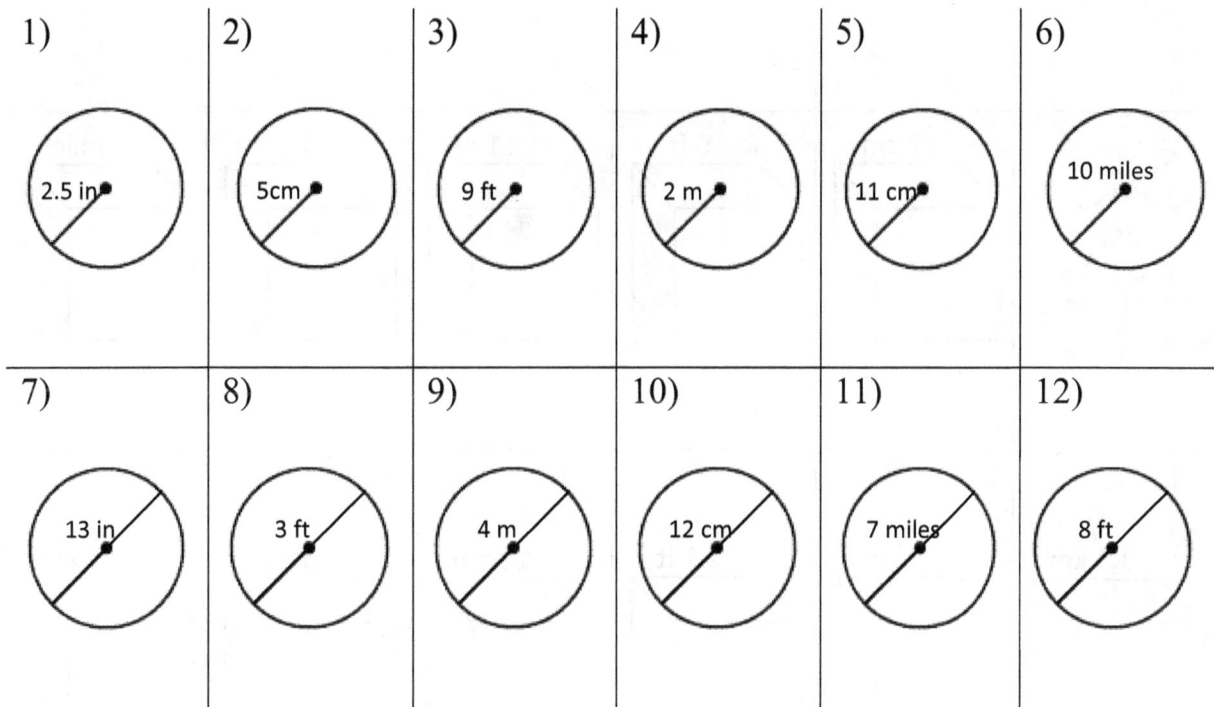

✎ **Complete the table below.** (π = 3.14)

Circle No.	Radius	Diameter	Circumference	Area
1	1 in	2 in	6.28 in	3.14 in^2
2		10 m		
3				28.26 ft^2
4			47.1 mi	
5		11 km		
6	7 cm			
7		12 ft		
8				314 m^2
9			56.52 in	
10	4.5 ft			

TASC Subject Test – Mathematics

Cubes

✎ Find the volume of each cube.

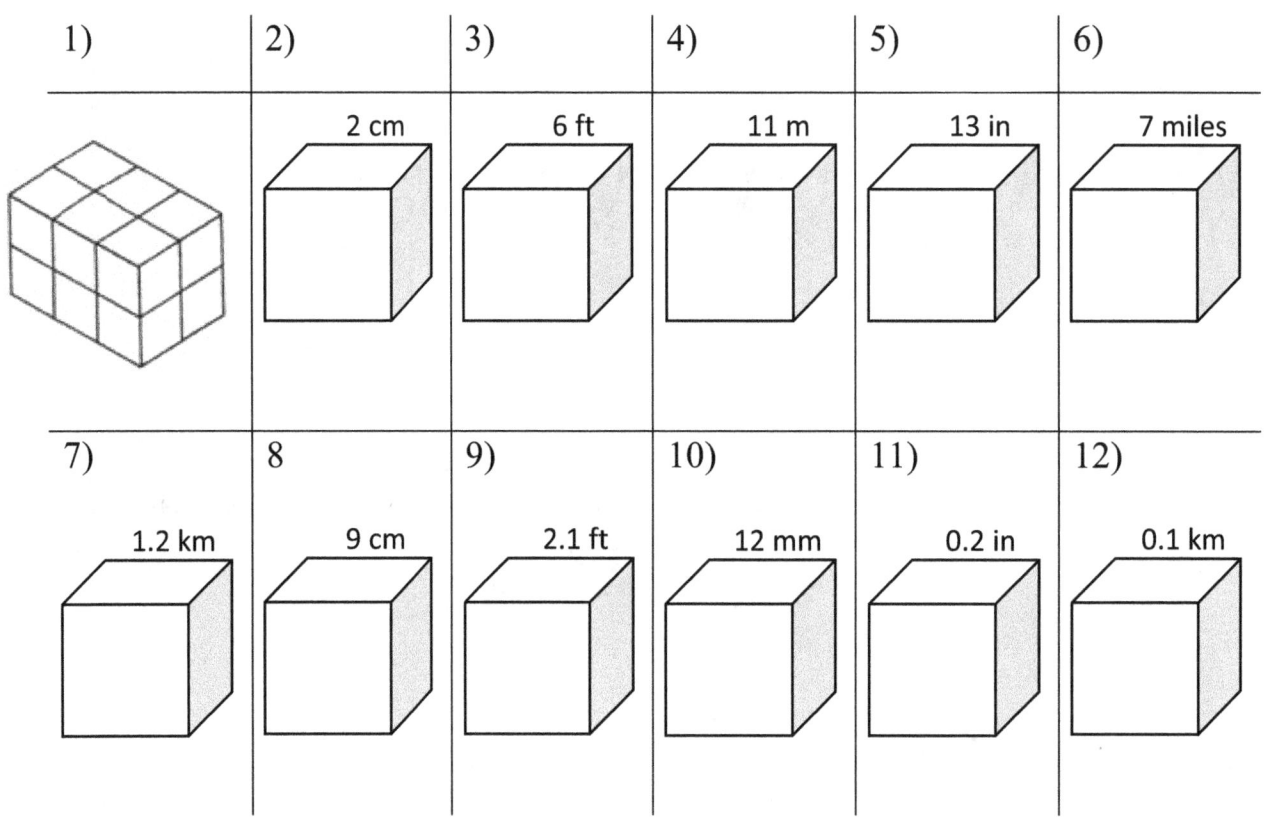

1)
2) 2 cm
3) 6 ft
4) 11 m
5) 13 in
6) 7 miles

7) 1.2 km
8) 9 cm
9) 2.1 ft
10) 12 mm
11) 0.2 in
12) 0.1 km

✎ Find the surface area of each cube.

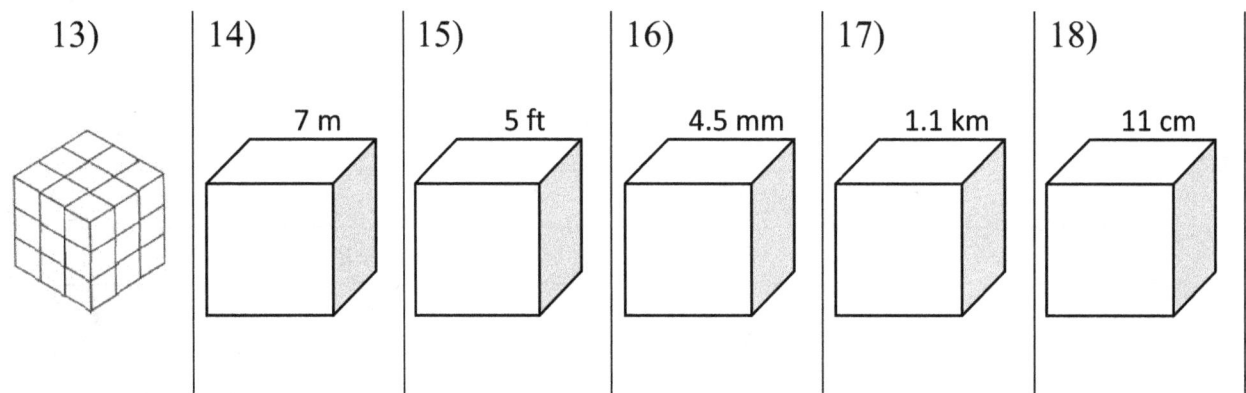

13)
14) 7 m
15) 5 ft
16) 4.5 mm
17) 1.1 km
18) 11 cm

TASC Subject Test – Mathematics

Rectangular Prism

✎ Find the volume of each Rectangular Prism.

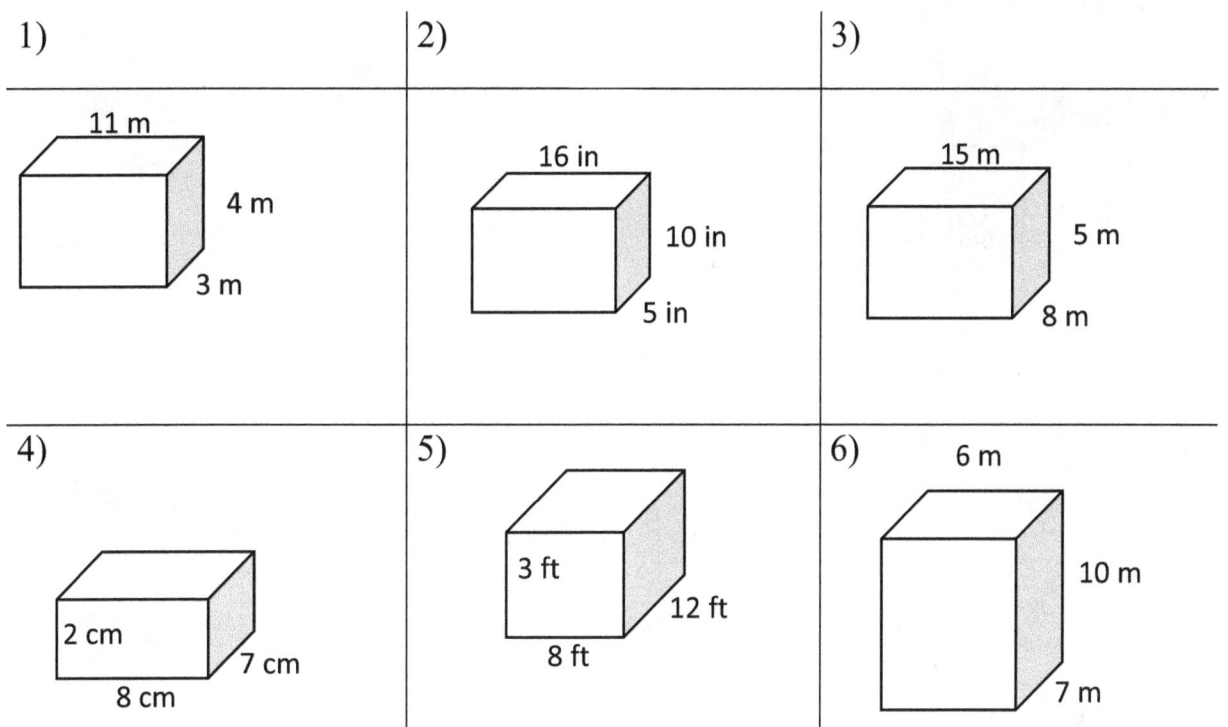

✎ Find the surface area of each Rectangular Prism.

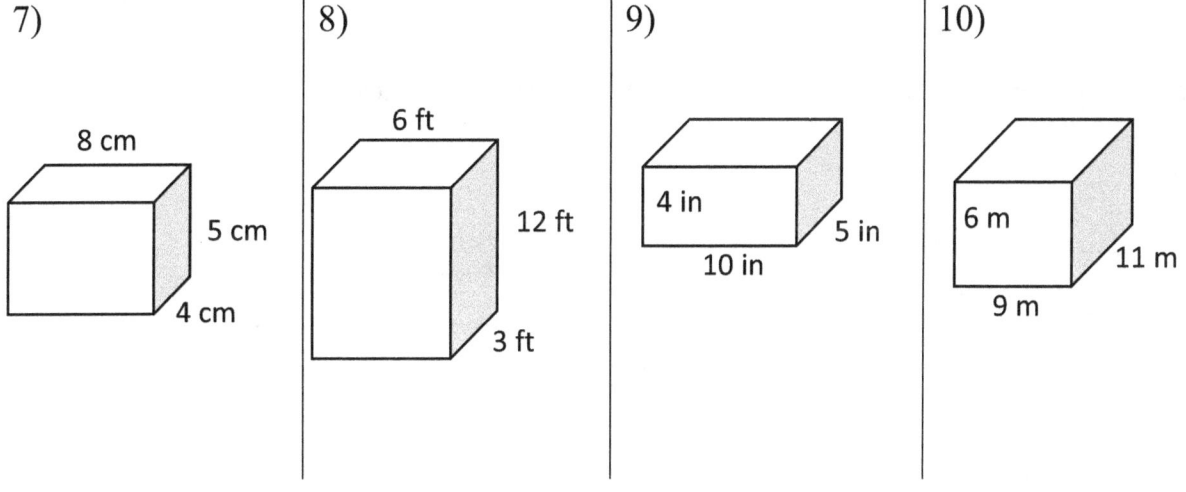

WWW.MathNotion.Com

131

TASC Subject Test – Mathematics

Cylinder

✏️ **Find the volume of each Cylinder. Round your answer to the nearest tenth.** ($\pi = 3.14$)

1)

2)

3)

4)

5)

6)

✏️ **Find the surface area of each Cylinder.** ($\pi = 3.14$)

7)

8)

9)

10)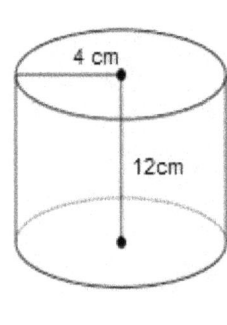

TASC Subject Test – Mathematics

Pyramids and Cone

✎ Find the volume of each Pyramid and Cone. ($\pi = 3.14$)

1)

2)

3)

4)

5)

6)

✎ Find the surface area of each Pyramid and Cone. ($\pi = 3.14$)

7)

8)

9)

10)

WWW.MathNotion.Com

TASC Subject Test – Mathematics

Answers of Worksheets

Angles

1) 16°   4) 34°   7) 90°   10) 75°
2) 96°   5) 70°   8) 75°   11) 70°
3) 59°   6) 52°   9) 33°

Pythagorean Relationship

1) No   5) Yes   9) 13   13) 15
2) Yes   6) No   10) 20   14) 30
3) No   7) Yes   11) 17   15) 36
4) Yes   8) Yes   12) 10   16) 12

Triangles

1) 60°   5) 45°   9) 54 square unites
2) 48°   6) 40°   10) 120 square unites
3) 55°   7) 45°   11) 90 square unites
4) 52°   8) 67°   12) 180 square unites

Polygons

1) 52 ft   5) 54 m   9) 30 m^2
2) 26 in   6) 25 cm   10) 300 in^2
3) 112 ft   7) 38 in   11) 160 km^2
4) 20 cm   8) 24 m   12) 49 in^2

Trapezoids

1) 50 cm^2 4) 60 cm^2 7) 36
2) 105 m^2 5) 80   8) 15
3) 39 ft^2 6) 24

Calculate

1) 13 cm   2) 11 ft   3) 12 m   4) 25 ft

Circles

1) 19.63 in^2 5) 379.94 cm^2 9) 12.56 m^2
2) 78.5 cm^2 6) 314 $miles^2$ 10) 113.04 cm^2
3) 254.34 ft^2 7) 132.67 in^2 11) 38.47 $miles^2$
4) 12.56 m^2 8) 7.07 ft^2 12) 50.24 ft^2

TASC Subject Test – Mathematics

Circle No.	Radius	Diameter	Circumference	Area
1	1 in	2 in	6.28 in	3.14 in^2
2	5 m	10 m	31.4 m	78.5 m^2
3	3 ft	6 ft	18.84 ft	28.26 ft^2
4	7.5 miles	15 mi	47.1 mi	176.63 mi^2
5	5.5 km	11 km	34.54 km	94.99 km^2
6	7 cm	14 cm	43.96 cm	153.86 cm^2
7	6 ft	12 ft	37.68 feet	113.04 ft^2
8	10 m	20 m	62.8 m	314 m^2
9	9 in	18 in	56.52 in	254.34 in^2
10	4.5 ft	9 ft	28.26 ft	63.585 ft^2

Cubes

1) 12
2) 8 cm^3
3) 216 ft^3
4) 1,331 m^3
5) 2,197 in^3
6) 343 $miles^3$
7) 1.728 km^3
8) 729 cm^3
9) 9.261 ft^3
10) 1,728 mm^3
11) 0.008 in^3
12) 0.001 km^3
13) 27
14) 294 m^2
15) 150 ft^2
16) 121.5 mm^2
17) 7.26 km^2
18) 726 cm^2

Rectangular Prism

1) 132 m^3
2) 800 in^3
3) 600 m^3
4) 112 cm^3
5) 288 ft^3
6) 420 m^3
7) 184 cm^2
8) 252 ft^2
9) 220 in^2
10) 438 m^2

Cylinder

1) 1,004.8 m^3
2) 214.6 cm^3
3) 9,495.4 cm^3
4) 1.1 m^3
5) 588.8 m^3
6) 452.2 in^3
7) 188.4 m^2
8) 602.9 cm^2
9) 37.7 cm^2
10) 401.9 m^2

Pyramids and Cone

1) 1,600 yd^3
2) 1,050 yd^3
3) 1,617 in^3
4) 392.5 m^3
5) 3,014.4 m^3
6) 366.33 cm^3
7) 1,440 yd^2
8) 1,536 m^2
9) 678.24 in^2
10) 1,205.76 cm^2

TASC Subject Test – Mathematics

TASC Subject Test – Mathematics

Chapter 11:
Statistics and Probability

Topics that you'll practice in this chapter:

- ✓ Mean and Median
- ✓ Mode and Range
- ✓ Histograms
- ✓ Stem–and–Leaf Plot
- ✓ Pie Graph
- ✓ Probability Problems

Mathematics is no more computation than typing is literature.

– John Allen Paulos

TASC Subject Test – Mathematics

Mean and Median

✎ **Find Mean and Median of the Given Data.**

1) 8, 7, 14, 4, 8

2) 14, 8, 25, 19, 16, 33, 11

3) 23, 18, 15, 12, 17

4) 34, 14, 10, 15, 6, 11

5) 10, 19, 6, 8, 32, 20, 17

6) 17, 26, 39, 69, 20, 6

7) 40, 38, 18, 11, 9, 2, 7, 32, 41

8) 24, 21, 31, 12, 33, 32, 22

9) 16, 14, 20, 41, 15, 20, 38, 4

10) 20, 20, 30, 18, 6, 28, 12, 46

11) 12, 7, 10, 11, 16, 22

12) 10, 29, 27, 12, 2, 15, 10, 3

✎ **Calculate.**

13) In a javelin throw competition, five athletics score 56, 34, 62, 23 and 19 meters. What are their Mean and Median? _____

14) Eva went to shop and bought 8 apples, 14 peaches, 6 bananas, 4 pineapples and 12 melons. What are the Mean and Median of her purchase? _____

15) Bob has 17 black pen, 19 red pen, 14 green pens, 20 blue pens and 5 boxes of yellow pens. If the Mean and Median are 19 respectively, what is the number of yellow pens in each box? _____

TASC Subject Test – Mathematics

Mode and Range

✎ **Find Mode and Rage of the Given Data.**

1) 4, 3, 7, 3, 3, 4
   Mode: _____   Range: _____

2) 18, 18, 24, 26, 18, 8, 14, 22
   Mode: _____   Range: _____

3) 8, 8, 8, 16, 19, 22, 20, 9, 13
   Mode: _____   Range: _____

4) 24, 24, 14, 28, 20, 18, 20, 24
   Mode: _____   Range: _____

5) 6, 21, 27, 24, 27, 27
   Mode: _____   Range: _____

6) 21, 8, 8, 7, 8, 12, 10, 22, 18, 13
   Mode: _____   Range: _____

7) 7, 4, 4, 6, 13, 13, 13, 0, 2, 2
   Mode: _____   Range: _____

8) 5, 8, 5, 14, 12, 14, 3, 5, 18
   Mode: _____   Range: _____

9) 7, 7, 7, 12, 7, 3, 8, 16, 3, 17
   Mode: _____   Range: _____

10) 15, 15, 19, 16, 4, 16, 10, 15
    Mode: _____   Range: _____

11) 6, 6, 5, 6, 42, 13, 19, 2
    Mode: _____   Range: _____

12) 8, 8, 9, 8, 9, 4, 34, 22
    Mode: _____   Range: _____

✎ **Calculate.**

13) A stationery sold 12 pencils, 56 red pens, 24 blue pens, 20 notebooks, 12 erasers, 21 rulers and 11 color pencils. What are the Mode and Range for the stationery sells?

    Mode: _____   Range: _____

14) In an English test, eight students score 10, 15, 15, 18 18, 16, 15 and 15. What are their Mode and Range? _____

15) What is the range of the first 6 even numbers greater than 8?
    _____

WWW.MathNotion.Com

TASC Subject Test – Mathematics

Times Series

✏ Use the following Graph to complete the table.

Day	Distance (km)
1	
2	

The following table shows the number of births in the US from 2007 to 2012 (in millions).

Year	Number of births (in millions)
2007	4.15
2008	3.70
2009	3.45
2010	3.20
2011	1.75
2012	2.98

Draw a Time Series for the table.

TASC Subject Test – Mathematics

Stem–and–Leaf Plot

✎ **Make stem ad leaf plots for the given data.**

1) 24, 26, 29, 20, 53, 27, 51, 55, 36, 21, 37, 30

2) 11, 59, 66, 14, 18, 19, 59, 65, 69, 61, 68, 65

3) 121, 55, 66, 54, 112, 128, 63, 125, 59, 123, 68, 119

4) 51, 32, 100, 56, 84, 36, 107, 56, 85, 39, 56, 106, 89

5) 33, 89, 19, 87, 81, 16, 11, 30, 86, 35, 17, 35, 13

6) 60, 92, 22, 25, 67, 93, 95, 62, 21, 64, 98, 29

TASC Subject Test – Mathematics

Pie Graph

The circle graph below shows all Robert's expenses for last month. Robert spent $140 on his hobbies last month.

Answer following questions based on the Pie graph.

1) How much was Robert's total expenses last month? _____

2) How much did Robert spend on his car last month? _____

3) How much did Robert spend for shopping last month? _____

4) How much did Robert spend on his rent last month? _____

5) What fraction is Robert's expenses for his rent and car out of his total expenses last month? _____

WWW.MathNotion.Com

TASC Subject Test – Mathematics

Probability Problems

✏ **Calculate.**

1) A number is chosen at random from 1 to 10. Find the probability of selecting number 6 or smaller numbers. _____

2) Bag A contains 18 red marbles and 6 green marbles. Bag B contains 16 black marbles and 8 orange marbles. What is the probability of selecting a green marble at random from bag A? What is the probability of selecting a black marble at random from Bag B? _____

3) A number is chosen at random from 1 to 20. What is the probability of selecting multiples of 4? _____

4) A card is chosen from a well-shuffled deck of 52 cards. What is the probability that the card will be a queen? _____

5) A number is chosen at random from 1 to 15. What is the probability of selecting a multiple of 3 or 5? _____

A spinner numbered 1–8, is spun once. What is the probability of spinning …?

6) an Odd number? _____   7) a multiple of 2? _____

8) a multiple of 5? _____   9) number 10? _____

TASC Subject Test – Mathematics

Answers of Worksheets

Mean and Median

1) Mean: 8.2, Median: 8
2) Mean: 18, Median: 16
3) Mean: 17, Median: 17
4) Mean: 15, Median: 12.5
5) Mean: 16, Median: 17
6) Mean: 29.5, Median: 23
7) Mean: 22, Median: 18
8) Mean: 25, Median: 24
9) Mean: 21, Median: 18
10) Mean: 22.5, Median: 20
11) Mean: 13, Median: 11.5
12) Mean: 13.5, Median: 11
13) Mean: 38.8, Median: 34
14) Mean: 8.8, Median: 8
15) 5

Mode and Range

1) Mode: 3, Range: 4
2) Mode: 18, Range: 18
3) Mode: 8, Range: 14
4) Mode: 24, Range: 14
5) Mode: 27, Range: 21
6) Mode: 8, Range: 15
7) Mode: 13, Range: 13
8) Mode: 5, Range: 15
9) Mode: 7, Range: 14
10) Mode: 15, Range: 15
11) Mode: 6, Range: 40
12) Mode: 8, Range: 30
13) Mode: 12, Range: 45
14) Mode: 15, Range: 8
15) 10

Time series

Day	Distance (km)
1	335
2	496
3	270
4	610
5	320
6	400

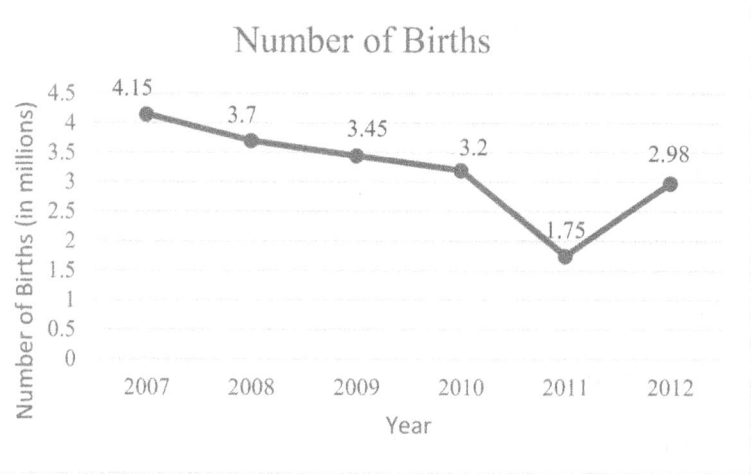

Stem–And–Leaf Plot

1)

Stem	leaf
2	0 1 4 6 7 9
3	0 6 7
5	1 3 5

2)

Stem	leaf
1	1 4 8 9
5	9 9
6	1 5 5 6 8 9

3)

Stem	leaf
5	4 5 9
6	3 6 8
11	2 9
12	1 3 5 8

TASC Subject Test – Mathematics

4)

Stem	leaf
3	2 6 9
5	1 6 6 6
8	4 5 9
10	0 6 7

5)

Stem	leaf
1	1 3 6 7 9
3	0 3 5 5
8	1 6 7 9

6)

Stem	leaf
2	2 1 5 9
6	0 2 4 7
9	2 3 5 8

Pie Graph

1) $5,600

2) $784

3) $1,288

4) $2,016

5) $\frac{1}{2}$

Probability Problems

1) $\frac{3}{5}$

2) $\frac{1}{4}, \frac{2}{3}$

3) $\frac{1}{4}$

4) $\frac{1}{13}$

5) $\frac{7}{15}$

6) $\frac{1}{2}$

7) $\frac{1}{2}$

8) $\frac{1}{8}$

9) 0

TASC Subject Test – Mathematics

Chapter 12 : TASC Test Review

The Test Assessing Secondary Completion, commonly known as the TASC or high school equivalency degree, is a standardized test. The TASC is a standardized test to verify that examinees have knowledge in core content areas equivalent to that of graduating high school seniors.

There are five subject area tests on TASC:
- Reading.
- Writing.
- Social Studies.
- Science.
- Mathematics.

The TASC Mathematics test is a 105-minute test that covers basic mathematics topics, quantitative problem-solving and algebraic questions. There are two Mathematics sections on the TASC. The first section contains 40 multiple choice questions where calculators are permitted. You have 50 minutes to complete this section. The second section contains 12 Gridded-Response questions. Calculator is NOT allowed in the second part. Test takers have 55 minutes to answer all questions in this section. Examinees will also be given a page of mathematic formulas to use during the test.

In this book, we have covered all Mathematics topics you need to know. Now, it's time to take a real TASC test. In this section, there are two complete TASC Mathematics Tests. Take these tests to see what score you'll be able to receive on a real TASC test.

TASC Subject Test – Mathematics

Time to Test

Time to refine your quantitative reasoning skill with a practice test

In this section, there are two complete TASCT Mathematics practice tests. Take these tests to simulate the test day experience. After you've finished, score your tests using the answer keys.

Before You Start

- You'll need a pencil, a calculator and a timer to take the test.
- For each question, there are four possible answers. Choose which one is best.
- It's okay to guess. There is no penalty for wrong answers.
- Use the answer sheet provided to record your answers.
- After you've finished the test, review the answer key to see where you went wrong.

Good Luck!

Mathematics is like love; a simple idea, but it can get complicated.

TASC Subject Test – Mathematics

TASC Practice Tests Answer Sheet

Remove (or photocopy) these answer sheets and use them to complete the practice tests.

TASC Practice Test – Section 1 Answer Sheet

#		#		#		#		#	
1	Ⓐ Ⓑ Ⓒ Ⓓ	9	Ⓐ Ⓑ Ⓒ Ⓓ	17	Ⓐ Ⓑ Ⓒ Ⓓ	25	Ⓐ Ⓑ Ⓒ Ⓓ	33	Ⓐ Ⓑ Ⓒ Ⓓ
2	Ⓐ Ⓑ Ⓒ Ⓓ	10	Ⓐ Ⓑ Ⓒ Ⓓ	18	Ⓐ Ⓑ Ⓒ Ⓓ	26	Ⓐ Ⓑ Ⓒ Ⓓ	34	Ⓐ Ⓑ Ⓒ Ⓓ
3	Ⓐ Ⓑ Ⓒ Ⓓ	11	Ⓐ Ⓑ Ⓒ Ⓓ	19	Ⓐ Ⓑ Ⓒ Ⓓ	27	Ⓐ Ⓑ Ⓒ Ⓓ	35	Ⓐ Ⓑ Ⓒ Ⓓ
4	Ⓐ Ⓑ Ⓒ Ⓓ	12	Ⓐ Ⓑ Ⓒ Ⓓ	20	Ⓐ Ⓑ Ⓒ Ⓓ	28	Ⓐ Ⓑ Ⓒ Ⓓ	36	Ⓐ Ⓑ Ⓒ Ⓓ
5	Ⓐ Ⓑ Ⓒ Ⓓ	13	Ⓐ Ⓑ Ⓒ Ⓓ	21	Ⓐ Ⓑ Ⓒ Ⓓ	29	Ⓐ Ⓑ Ⓒ Ⓓ	37	Ⓐ Ⓑ Ⓒ Ⓓ
6	Ⓐ Ⓑ Ⓒ Ⓓ	14	Ⓐ Ⓑ Ⓒ Ⓓ	22	Ⓐ Ⓑ Ⓒ Ⓓ	30	Ⓐ Ⓑ Ⓒ Ⓓ	38	Ⓐ Ⓑ Ⓒ Ⓓ
7	Ⓐ Ⓑ Ⓒ Ⓓ	15	Ⓐ Ⓑ Ⓒ Ⓓ	23	Ⓐ Ⓑ Ⓒ Ⓓ	31	Ⓐ Ⓑ Ⓒ Ⓓ	39	Ⓐ Ⓑ Ⓒ Ⓓ
8	Ⓐ Ⓑ Ⓒ Ⓓ	16	Ⓐ Ⓑ Ⓒ Ⓓ	24	Ⓐ Ⓑ Ⓒ Ⓓ	32	Ⓐ Ⓑ Ⓒ Ⓓ	40	Ⓐ Ⓑ Ⓒ Ⓓ

TASC Practice Test: Section 2: Grid-ins Questions

41, 42, 43, 44 (grid-in answer bubbles 0–9)

WWW.MathNotion.Com

TASC Subject Test – Mathematics

TASC Subject Test – Mathematics

Mathematics Reference Sheet

Cylinder: $v = \pi r^2 h$

Pyramid: $v = \frac{1}{3} bh$

cone: $v = \frac{1}{3} \pi r^2 h$

Sphere: $\frac{4}{3} \pi r^3$

coordinate Geometry

Midpoint of the segment AB:

$M\left(\frac{x_1+x_2}{2}, \frac{y_1+y_2}{2}\right)$

Distance from A to B:

$d = \sqrt{(x_1 - x_2)^2 + (y_1 - y_2)^2}$

Slope of a line:

$m = \frac{y_2 - y_1}{x_2 - x_1} = \frac{rise}{run}$

Special Factoring

$a^2 - b^2 = (a+b)(a-b)$

$a^2 + 2ab + b^2 = (a+b)(a+b)$

$a^2 - 2ab + b^2 = (a-b)(a-b)$

$a^3 + b^3 = (a+b)(a^2 - ab + b^2)$

$a^3 - b^3 = (a-b)(a^2 + ab + b^2)$

Quadratic Formula

for $ax^2 + bx + c = 0$

$x = \frac{-b \pm \sqrt{b^2 - 4ac}}{2a}$

Interest

Simple Interest:

$I = prt$

Interest Formula (compounded n times per year):

$A = p\left(1 + \frac{r}{n}\right)^{nt}$

A = Amount after t years.

p = principal

r = annual interest rate

t = time in years

I = Interest

Trigonometric Identities

Pythagorean Theorem:

$a^2 + b^2 = c^2$

$\sin \theta = \frac{opp}{hyp}$

$\cos \theta = \frac{adj}{hyp}$

$\tan \theta = \frac{opp}{adj}$

$\sin^2 \theta + \cos^2 \theta = 1$

Density $= \frac{Mass}{Volume}$

TASC Subject Test – Mathematics

Central Angle	Inscribed Angle	Intersecting Chords Theorem
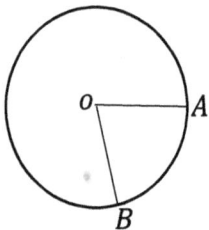 $m\angle AOB = m\widehat{AB}$	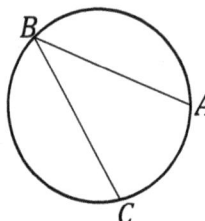 $m\angle ABC = \frac{1}{2} m\widehat{AC}$	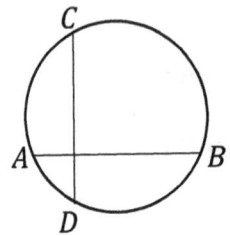 $A \cdot B = C \cdot D$

Probability

Permutations: $_nP_r = \dfrac{n!}{(n-r)!}$

Combinations: $_nC_r = \dfrac{n!}{(n-r)!\, r!}$

Multiplication rule (independent events): $P(A \text{ and } B) = P(A) \cdot P(B)$

Multiplication rule (general): $P(A \text{ and } B) = P(A) \cdot P(B|A)$

Addition rule: $P(A \text{ or } B) = P(A) + P(B) - P(A \text{ and } B)$

Conditional Probability: $P(B|A) = \dfrac{P(A \text{ and } B)}{P(A)}$

Arithmetic Sequence: $a_n = a_1 + (n-1)d$ where a_n is the nth term, a_1 is the first term, and d is the common difference.

Geometric Sequence: $a_n = a_1 r^{(n-1)}$ where a_n is the nth term, a_1 is the first term, and r is the common ratio.

TASC Subject Test – Mathematics

TASC Practice Test 1

Mathematics

Section 1 - Calculator

- ✓ **40 questions**
- ✓ **Total time for this section: 50 Minutes**
- ✓ **You may use a calculator on this Section.**

Administered *Month Year*

TASC Subject Test – Mathematics

1) 63 is What percent of 42?

    A. 150 %

    B. 50 %

    C. 15 %

    D. 115 %

2) The marked price of a computer is E Euro. Its price decreased by 30% in March and later increased by 20% in April. What is the final price of the computer in E Euro?

    A. 0.56 E

    B. 1.04 E

    C. 0.84 E

    D. 0.90 E

3) A rope weighs 720 grams per meter of length. What is the weight in kilograms of 13.5 meters of this rope? (1 kilograms = 1,000 grams)

    A. 0.972

    B. 9.72

    C. 97.20

    D. 972.00

4) Which of the following could be the product of two consecutive prime numbers? (Select one or more answer choices)

    A. 133

    B. 77

    C. 144

    D. 66

5) A $72 shirt now selling for $18 is discounted by what percent?

    A. 70 %

    B. 25 %

    C. 30 %

    D. 75 %

WWW.MathNotion.Com

TASC Subject Test – Mathematics

6) The score of Zoe was one-seventh of Emma and the score of Harper was twice that of Emma. If the score of Harper was 84, what is the score of Zoe?

   A. 6

   B. 24

   C. 8

   D. 12

7) Five seventh of 63 is equal to $\frac{5}{9}$ of what number?

   A. 18

   B. 198

   C. 81

   D. 189

8) A bag contains 25 balls: seven green, three black, three blue, seven red and five white. If 15 balls are removed from the bag at random, what is the probability that a white ball has been removed?

   A. $\frac{3}{5}$

   B. $\frac{1}{25}$

   C. $\frac{4}{25}$

   D. $\frac{1}{5}$

9) What is the value of 8^3?

   A. 64

   B. 512

   C. 256

   D. 1,024

10) What is the median of these numbers? 15, 8, 19, 31, 11, 11, 10, 15, 27

   A. 13

   B. 19

   C. 11

   D. 15

TASC Subject Test – Mathematics

11) How many tiles of 6 cm² is needed to cover a floor of dimension 6 cm by 17 cm?

    A. 17   C. 34

    B. 18   D. 51

12) Ryan traveled 400 km in 8 hours and Riley traveled 720 km in 9 hours. What is the ratio of the average speed of Ryan to average speed of Riley?

    A. 3: 5   C. 4: 9

    B. 8: 9   D. 5: 8

13) An angle is equal to one fifth of its supplement. What is the measure of that angle?

    A. 30   C. 50

    B. 45   D. 36

14) Abigail purchased a sofa for $349.20 The sofa is regularly priced at $485. What was the percent discount Abigail received on the sofa?

    A. 26%   C. 28%

    B. 72%   D. 38%

15) Find the average of the following numbers: 38, 24, 32 and 20?

    A. 22.5   C. 30

    B. 28.5   D. 26

16) When a number is subtracted from 56 and the difference is divided by that number, the result is 7. What is the value of the number?

    A. 6   C. 9

    B. 7   D. 14

WWW.MathNotion.Com

TASC Subject Test – Mathematics

17) Right triangle ABC has two legs of lengths 10 cm (AB) and 24 cm (AC). What is the length of the third side (BC)?

   A. 34 cm   C. 26 cm

   B. 40 cm   D. 25 cm

18) A taxi driver earns $19 per hour work. If he works 10 hours a day, and he uses 7-liters Petrol in 2 hours with price $2.5 for 1-liter. How much money does he earn in one day?

   A. $102.50   C. $124.50

   B. $112.50   D. $122.50

19) The width of a box is one fifth of its length. The height of the box is one tenth of its width. If the length of the box is 100 cm, what is the volume of the box?

   A. 200 cm^3 C. 400 cm^3

   B. 2,000 cm^3 D. 4,000 cm^3

20) The price of a sofa is decreased by 30% to $315. What was its original price?

   A. $540   C. $460

   B. $450   D. $640

21) If 52% of a class are girls, and 35% of girls play tennis, what percent of the class play tennis?

   A. 182%   C. 18.2%

   B. 224%   D. 22.4%

WWW.MathNotion.Com

TASC Subject Test – Mathematics

22) The price of a car was $90,000 in 2014, $72,000 in 2015 and $57,600 in 2016. What is the rate of depreciation of the price of car per year?

   A. 15 %   C. 20 %

   B. 22 %   D. 25 %

23) The average of five consecutive numbers is 112. What is the smallest number?

   A. 102   C. 120

   B. 115   D. 110

24) The area of a circle is less than 144π. Which of the following can be the circumference of the circle? (Select one or more answer choices)

   A. 28π C. 22π

   B. 32π D. 25π

25) A bank is offering 2.25% simple interest on a savings account. If you deposit $10,000, how much interest will you earn in four years?

   A. $450   C. $900

   B. $10,450   D. $10,900

26) In three successive hours, a car travels 70 km, 77 km, and 80 km. In the next five hours, it travels with an average speed of 65 km per hour. Find the total distance the car traveled in 8 hours.

   A. 552 km   C. 325 km

   B. 227 km   D. 642 km

WWW.MathNotion.Com

TASC Subject Test – Mathematics

27) The ratio of boys to girls in a school is 9: 5. If there are 490 students in a school, how many boys are in the school.

   A. 230

   B. 330

   C. 315

   D. 35

28) If the area of trapezoid is 210 cm^2, what is the perimeter of the trapezoid?

   A. 55

   B. 64

   C. 50

   D. 60

29) In the xy-plane, the point $(-1, -4)$ and $(2, 5)$ are on the line A. Which of the following points could also be on the line A?

   A. $(2, 9)$

   B. $(-2, 5)$

   C. $(4, 11)$

   D. $(0, 2)$

30) A chemical solution contains 12% alcohol. If there is 21.6 ml of alcohol, what is the volume of the solution?

   A. 190 ml

   B. 170 ml

   C. 150 ml

   D. 180 ml

TASC Subject Test – Mathematics

31) The average high of 15 constructions in a town is 140 m, and the average high of 5 towers in the same town is 120 m. What is the average high of all the 20 structures in that town?

   A. 152

   B. 135

   C. 125

   D. 153

32) The price of a laptop is decreased by 15% to $170. What is its original price?

   A. 1,133

   B. 195.50

   C. 200

   D. 220

33) In 1999, the average worker's income increased $1,500 per year starting from $22,700 annual salary. Which equation represents income greater than average? (I = income, x = number of years after 1999)

   A. $I > 1,500\,x + 22,700$

   B. $I > -1,500\,x + 22,700$

   C. $I < 22,700\,x + 1,500$

   D. $I < -22,700\,x + 1,500$

34) A boat sails 20 miles south and then 15 miles east. How far is the boat from its start point?

   A. 150 miles

   B. 60 miles

   C. 35 miles

   D. 25 miles

35) If 64% of F is 32% of M, then F is what percent of M?

   A. 80 %

   B. 20 %

   C. 50 %

   D. 200 %

WWW.MathNotion.Com

TASC Subject Test – Mathematics

36) How many possible outfit combinations come from nine shirts, five slacks, and 7 times?

   A. 153

   B. 315

   C. 45!

   D. 35!

37) The surface area of a cylinder is $72\pi \ cm^2$. If its height is 5 cm, what is the radius of the cylinder?

   A. 2 cm

   B. 9 cm

   C. 4 cm

   D. 5 cm

38) How long does a 338–miles trip take moving at 52 miles per hour (mph)?

   A. 6 hours

   B. 6 hours and 30 minutes

   C. 5 hours and 36 minutes

   D. 6 hours and 45 minutes

39) A shirt costing $500 is discounted 30%. After a month, the shirt is discounted another 10%. Which of the following expressions can be used to find the selling price of the shirt?

   A. $(500) - 500(0.70)$

   B. $(500)(0.70)(0.90)$

   C. $(500)(0.30) - (150)(0.10)$

   D. $(500)(0.30) - (500)(0.10)$

40) Which of the following lists shows the fractions in order from least to greatest?

   A. $\frac{1}{8}, \frac{3}{14}, \frac{2}{9}$

   B. $\frac{3}{14}, \frac{1}{8}, \frac{2}{9}$

   C. $\frac{2}{9}, \frac{1}{8}, \frac{3}{14}$

   D. $\frac{1}{8}, \frac{2}{9}, \frac{3}{14}$

TASC Subject Test – Mathematics

IF YOU FINISH BEFORE TIME IS CALLED, YOU MAY CHECK YOUR WORK ON THIS SECTION ONLY. DO NOT TURN TO OTHER SECTION IN THE TEST. **STOP**

TASC Subject Test – Mathematics

TASC Practice Test 1

Mathematics

Section 2 - No Calculator

- ✓ 12 questions
- ✓ Total time for this section: 55 Minutes
- ✓ You may NOT use a calculator on this Section.

Administered *Month Year*

TASC Subject Test – Mathematics

41) From last year, the price of gasoline has increased from $2.25 per gallon to $3.15 per gallon. The new price is what percent of the original price?

42) If $5x - 9 = 6$, What is the value of $4x - \frac{3}{4}$?

43) What is the area of an isosceles right triangle that has one leg that measures 14 cm?

44) What is the value of $f(3)$ for the following function f?

$$f(x) = 3x^2 - 7x + 12$$

45) If the ratio of $4a$ to $9b$ is $\frac{1}{15}$, what is the ratio of a to b?

TASC Subject Test – Mathematics

46) A construction company is building a wall. The company can build 14 cm of the wall per minute. After 50 minutes $\frac{7}{9}$ of the wall is completed. How many meters is the wall?

47) $8 \times (-5) + 10 - 5(-4 - 14 \times 6) \div 11 = ?$

48) The perimeter of the trapezoid below is 52. What is its area?

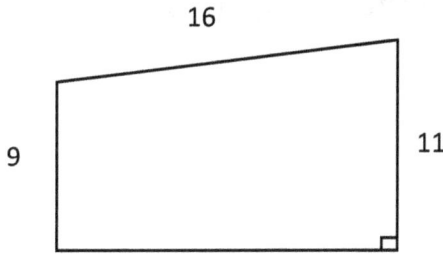

49) The sum of two numbers is M. if one of the numbers is 6, then twice of the other number would be what?

TASC Subject Test – Mathematics

50) The volume of cube A is $\frac{1}{3}$ of its surface area. What is the length of an edge of cube A?

51) If $\frac{3x-3}{4} = N$ and $N = -6$, what is the value of x?

52) A ladder leans against a wall forming a 45° angle between the ground and the ladder. If the bottom of the ladder is 15 feet away from the wall, how many feet is the ladder?

TASC Subject Test – Mathematics

TASC Practice Test 2

Mathematics

Section 1 - Calculator

- ✓ 40 questions
- ✓ Total time for this section: 50 Minutes
- ✓ You may use a calculator on this Section.

Administered *Month Year*

TASC Subject Test – Mathematics

1) What is the perimeter of a square in centimeters that has an area of 547.56 cm²?

   A. 23.4

   B. 93.6

   C. 96.3

   D. 103.6

2) If 57% of A is 19% of B, then B is what percent of A?

   A. 33%

   B. 66%

   C. 3%

   D. 300%

3) Simplify the expression.
$$(8x^4 - 4x^2 - 3x^5) - (3x^2 - 2x^5 + 5x^4)$$

   A. $-(x^5 + 3x^3 - 7x^2)$

   B. $2(x^5 + 2x^4 - 4x^2)$

   C. $-(x^5 - 3x^4 + 7x^2)$

   D. $2(-x^5 - 2x^4 - 4x^2)$

4) Mr. Matthews saves $2,400 out of his annually family income of $55,200. What fractional part of his income does he save?

   A. $\frac{1}{23}$

   B. $\frac{8}{23}$

   C. $\frac{7}{23}$

   D. $\frac{3}{23}$

5) What is the median of these numbers? 25, 9, 4, 34, 28, 17, 51

   A. 51

   B. 11.5

   C. 34

   D. 25

6) A bank is offering 1.65% simple interest on a savings account. If you deposit $16,000, how much interest will you earn in four years?

   A. $1,650

   B. $1,056

   C. $2,106

   D. $10,056

TASC Subject Test – Mathematics

7) Last week 22,000 fans attended a football match. This week three times as many bought tickets, but one sixth of them cancelled their tickets. How many are attending this week?

    A. 77,000   C. 66,000

    B. 11,000   D. 55,000

8) In two successive years, the population of a town is increased by 6% and 15%. What percent of the population is increased after two years?

    A. 29.9%   C. 12.19 %

    B. 21.9%   D. 29.1 %

9) The mean of 30 test scores was calculated as 35. But it turned out that one of the scores was misread as 35 but it was 20. What is the correct mean of the test scores?

    A. 34   C. 34.5

    B. 35   D. 36.5

10) Which of the following graphs represents the compound inequality $-2 \leq 5x - 7 < 3$?

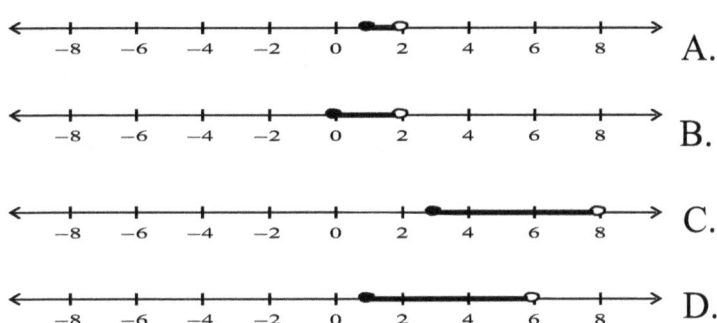

TASC Subject Test – Mathematics

11) Two dice are thrown simultaneously, what is the probability of getting a sum of 6 or 9?

   A. $\frac{1}{5}$

   B. $\frac{5}{18}$

   C. $\frac{1}{9}$

   D. $\frac{1}{4}$

12) Which of the following shows the numbers in descending order?

$$\frac{1}{8}, 0.4, 9\%, \frac{1}{12}$$

   A. $\frac{1}{8}, 9\%, \frac{1}{12}, 0.4$

   B. $9\%, \frac{1}{8}, 0.4, \frac{1}{12}$

   C. $0.4, \frac{1}{8}, 9\%, \frac{1}{12}$

   D. $0.4, 9\%, \frac{1}{8}, \frac{1}{12}$

13) What is the area of a square whose diagonal is 8?

   A. 64

   B. 32

   C. 16

   D. 48

14) What is the volume of a box with the following dimensions?

   Height = 11 cm   Width = 5 cm   Length = 3 cm

   A. 165 cm^3

   B. 256 cm^3

   C. 156 cm^3

   D. 55 cm^3

15) The average of 5 numbers is 29. The average of 3 of those numbers is 45. What is the average of the other two numbers?

   A. 16

   B. 10

   C. 5

   D. 15

WWW.MathNotion.Com

TASC Subject Test – Mathematics

16) What is the value of y in the following system of equations?

$$4x + 5y = -11$$
$$3x - 2y = 9$$

A. 3

B. 5

C. −5

D. −3

17) The perimeter of a rectangular yard is 32 meters. What is its length if its width is triple its length?

A. 4 meters

B. 12 meters

C. 9 meters

D. 20 meters

18) In a stadium the ratio of home fans to visiting fans in a crowd is 4:7. Which of the following could be the total number of fans in the stadium? (Select one or more answer choices)

A. 38,980

B. 47,300

C. 41,500

D. 53,670

19) What is the equivalent temperature of $86°F$ in Celsius? $C = \frac{5}{9}(F - 32)$

A. 32

B. 38

C. 30

D. 35

20) Which of the following points lies on the line $3x - 2y = -8$? (Select one or more answer choices)

A. $(-1, 3)$

B. $(2, -2)$

C. $(0, 4)$

D. $(5, 1)$

TASC Subject Test – Mathematics

21) Mr. Jefferson family are choosing a menu for their reception. They have 4 choices of appetizers, 10 choices of entrees, 6 choices of cake. How many different menu combinations are possible for them to choose?

　A. 24　　　　　　　　　　　C. 80

　B. 34　　　　　　　　　　　D. 240

22) Anita's trick–or–treat bag contains 32 pieces of chocolate, 25 suckers, 15 pieces of gum, 28 pieces of licorice. If she randomly pulls a piece of candy from her bag, what is the probability of her pulling out a piece of gum?

　A. $\dfrac{3}{40}$　　　　　　　　　　C. $\dfrac{1}{102}$

　B. $\dfrac{1}{24}$　　　　　　　　　　D. $\dfrac{3}{20}$

23) The average of five numbers is 42. If a sixth number that is greater than 60 is added, then, which of the following could be the new average? (Select one or more answer choices)

　A. 46　　　　　　　　　　　C. 38

　B. 42　　　　　　　　　　　D. 44

24) The ratio of boys and girls in a class is 2:5. If there are 63 students in the class, how many more boys should be enrolled to make the ratio 1:1?

　A. 27　　　　　　　　　　　C. 18

　B. 7　　　　　　　　　　　　D. 24

TASC Subject Test – Mathematics

25) The perimeter of the trapezoid below is 63 cm. What is its area?

A. 310 cm²

B. 216 cm²

C. 360 cm²

D. 266 cm²

26) A card is drawn at random from a standard 52–card deck, what is the probability that the card is of spades? (The deck includes 13 of each suit clubs, diamonds, hearts, and spades)

A. $\frac{1}{13}$

B. $\frac{1}{4}$

C. $\frac{1}{52}$

D. $\frac{1}{2}$

27) What is the surface area of the cylinder below?

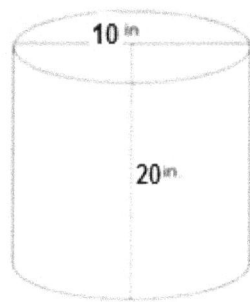

A. 500π in²

B. 300π in²

C. 260π in²

D. 250π in²

28) Simplify $2x^3y^6(-5xy^4)^2$.

A. $10x^4y^{12}$

B. $-10x^4y^{12}$

C. $50x^5y^{14}$

D. $-50x^5y^{14}$

TASC Subject Test – Mathematics

29) Daniel is 24 miles ahead of Noa and running at 5.2 miles per hour. Noa is running at the speed of 4 miles per hour. How long does it take Noa to catch Daniel?

A. 20 hours, and 15 minutes

B. 10 hours, 50 minutes

C. 20 hours

D. 10 hours

30) A football team had $35,000 to spend on supplies. The team spent $18,000 on new balls. New sport shoes cost $98 each. Which of the following inequalities represent the number of new shoes the team can purchase?

A. $98x - 18,000 \leq 35,000$

B. $98x + 18,000 \leq 35,000$

C. $17,000 + 98x \geq 35,000$

D. $17,000 - 98x \geq 35,000$

31) If 70% of a number is 35, what is the number?

A. 5

B. 18

C. 40

D. 50

32) The square of a number is $\frac{48}{243}$. What is the cube of that number?

A. $\frac{121}{289}$

B. $\frac{64}{729}$

C. $\frac{36}{49}$

D. $\frac{16}{81}$

33) 80 students took an exam and 24 of them failed. What percent of the students passed the exam?

A. 35%

B. 25%

C. 70%

D. 30%

TASC Subject Test – Mathematics

34) The circle graph below shows all Mr. Wilson's expenses for last month. If he spent $1,200 on his car, how much did he spend for his rent?

A. $1,000

B. $400

C. $800

D. $1,200

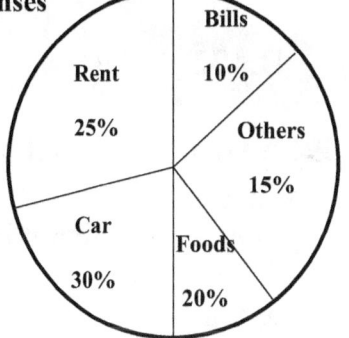

Mr. Wilson's monthly expenses

35) What is the value of x in the following equation?

$$\frac{2}{5}x + \frac{1}{24} = \frac{1}{8}$$

A. $\frac{4}{25}$

B. $\frac{1}{12}$

C. $\frac{5}{24}$

D. $\frac{24}{5}$

36) If 140% of a number is 210, then what is the 40% of that number?

A. 35

B. 85

C. 60

D. 75

37) Mrs. Thomson needs an 82% average in her writing class to pass. On her first 4 exams, he earned scores of 72%, 76%, 85%, and 93%. What is the minimum score Mrs. Thomson can earn on her fifth and final test to pass?

A. 84

B. 83

C. 93.7

D. 84.7

TASC Subject Test – Mathematics

38) The length of a rectangle is 2 meters less than 3 times its width. The perimeter of the rectangle is 52 meters. What is the area of the rectangle in meters?

   A. 196   C. 182

   B. 172   D. 130

39) A swimming pool holds 4,320 cubic feet of water. The swimming pool is 24 feet long and 15 feet wide. How deep is the swimming pool?

   A. 2.4   C. 24

   B. 12   D. 60

40) Right triangle ABC has two legs of lengths 12 cm (AB) and 16 cm (AC). What is the length of the third side (BC)?

   A. 26 cm   C. 20 cm

   B. 24 cm   D. 22 cm

IF YOU FINISH BEFORE TIME IS CALLED, YOU MAY CHECK YOUR WORK ON THIS SECTION ONLY. DO NOT TURN TO OTHER SECTION IN THE TEST. **STOP**

TASC Subject Test – Mathematics

TASC Practice Test 2

Mathematics

Section 2 - No Calculator

- ✓ 12 questions
- ✓ Total time for this section: 55 Minutes
- ✓ You may NOT use a calculator on this Section.

Administered Month Year

TASC Subject Test – Mathematics

41) A tree 70 feet tall casts a shadow 25 feet long. Jack is 7 feet tall. How long is Jack's shadow?

42) What is the value of the expression $5(2x + 3y) + (14 - 5x)^2$ when $x = 2$ and $y = -1$?

43) What is the slope of a line that is perpendicular to the line $x - 8y = 9$?

44) $[4 \times (-16)] + ([(-5) \times (-16)] \div 5) + (-23) = ?$

45) What is the product of all possible values of x in the following equation?

$$|-4x + 2| = 30$$

46) What is the value of x in the following equation? $-35 = 81 - x$

TASC Subject Test – Mathematics

47) If $7x - 25 = -4$, what is the value of $3x - \frac{1}{2}$?

48) The width of a box is one fourth of its length. The height of the box is one fourth of its width. If the length of the box is 48 cm, what is the volume of the box?

49) The area of a rectangular yard is 224 square meters. What is its width if its length is 16 meters?

50) In a classroom of 60 students, 15 are female. What percentage of the class is male?

51) One ninth of 135 is equal to $\frac{3}{7}$ of what number?

TASC Subject Test – Mathematics

52) The average weight of 27 girls in a class is 38 kg and the average weight of 23 boys in the same class is 52 kg. What is the average weight of all the 50 students in that class?

IF YOU FINISH BEFORE TIME IS CALLED, YOU MAY CHECK YOUR WORK ON THIS SECTION ONLY. DO NOT TURN TO OTHER SECTION IN THE TEST. **STOP**

TASC Subject Test – Mathematics

Chapter 13 : Answers and Explanations

Answer Key

❋ Now, it's time to review your results to see where you went wrong and what areas you need to improve!

TASC Mathematics Practice Tests

TASC Practice Test 1

Section 1		Section 2			
1	A	16	B	31	B	41	140
2	C	17	C	32	C	42	11.25
3	B	18	A	33	A	43	98
4	B	19	D	34	D	44	18
5	D	20	B	35	D	45	0.15
6	A	21	C	36	B	46	9
7	C	22	C	37	C	47	10
8	D	23	D	38	B	48	160
9	B	24	C	39	B	49	2M−12
10	D	25	C	40	A	50	2
11	A	26	A			51	−7
12	D	27	C			52	$15\sqrt{2}$
13	A	28	D				
14	C	29	C				
15	B	30	D				

TASC Practice Test 2

Section 1		Section 2			
1	B	16	D	31	D	41	2.5
2	D	17	A	32	B	42	21
3	C	18	B	33	C	43	−8
4	A	19	C	34	A	44	−71
5	D	20	C	35	C	45	−56
6	B	21	D	36	C	46	116
7	D	22	D	37	A	47	8.5
8	B	23	A	38	C	48	1,728
9	C	24	A	39	B	49	14
10	A	25	B	40	D	50	75
11	D	26	B			51	35
12	C	27	D			52	44.44
13	B	28	C				
14	A	29	C				
15	C	30	B				

TASC Subject Test – Mathematics

TASC Subject Test – Mathematics

Practice Test 1

Section 1- Calculator

1) Answer: A

Use percent formula: $Part = \frac{percent \times whole}{100}$

$63 = \frac{percent \times 42}{100} \Rightarrow \frac{63}{1} = \frac{percent \times 42}{100}$, cross multiply.

$6,300 = percent \times 42$, divide both sides by 42: $150 = percent$

2) Answer: C

To find the discount, multiply the number by (100% − rate of discount).

Therefore, for the first discount we get: $(100\% - 30\%)(E) = (0.70)E$

For increase of 20%: $(0.70)E \times (100\% + 20\%) = (0.70)(1.20) = 0.84E$.

3) Answer: B

The weight of 13.5 meters of this rope is: $13.5 \times 720g = 9,720g$

1 kg = 1,000 g, therefore, $9,720 \, g \div 1000 = 9.72 kg$

4) Answer: B

Some of prime numbers are: 2, 3, 5, 7, 11, 13, 17, 19

Find the product of two consecutive prime numbers:

5 × 7 = 35 (not in the options!)

7 × 11 = 77 (bingo!)

11 × 13 = 143 (not in the options)

Choice B is correct.

5) Answer: D

Use the formula for Percent of Change $= \frac{New\ Value - Old\ Value}{Old\ Value} \times 100\%$

$\frac{72-18}{72} \times 100\% = -75\%$ (negative sign here means that the new price is less than old price).

6) Answer: A

If the score of Harper was 84, therefore the score of Emma is 42. Since, the score of Zoe was one seventh of Emma, therefore, the score of Zoe is 6.

WWW.MathNotion.Com

TASC Subject Test – Mathematics

7) Answer: C

Let x be the number. Write the equation and solve for x.

$\frac{5}{7} \times 63 = \frac{5}{9} \cdot x \Rightarrow \frac{5 \times 63}{7} = \frac{5x}{9}$, use cross multiplication to solve for x.

$45 \times 63 = 5x \times 7 \Rightarrow 2{,}835 = 35x \Rightarrow x = 81$

8) Answer: D

If 15 balls are removed from the bag at random, there will be five ball in the bag.

The probability of choosing a white ball is 5 out of 25. Therefore, the probability of not choosing a white ball is 15 out of 25 and the probability of having not a white ball after removing 15 balls is the same.

9) Answer: B

$8^3 = 8 \times 8 \times 8 = 512$

10) Answer: D

Write the numbers in order:

8, 10, 11, 11, 15, 15, 19, 27, 31

Since we have 9 numbers (9 is odd), then the median is the number in the middle, which is 15.

11) Answer: A

The area of the floor is: $6\ cm \times 17\ cm = 102\ cm^2$

The number of tiles needed $= 102 \div 6 = 17$

12) Answer: D

The average speed of Ryan is: $400 \div 8 = 50\ km$

The average speed of Riley is: $720 \div 9 = 80\ km$

Write the ratio and simplify. $50 : 80 \Rightarrow 5 : 8$.

13) Answer: A

The sum of supplement angles is 180. Let x be that angle. Therefore,

$x + 5x = 180$.

$6x = 180$, divide both sides by 6: $x = 30$.

TASC Subject Test – Mathematics

14) Answer: C

Use percent formula: $Part = \frac{percent \times whole}{100}$

$349.20 = \frac{percent \times 485}{100} \Rightarrow$ (cross multiply): $34,920 = percent \times 485 \Rightarrow$ $percent = \frac{34,920}{485} = 72$

349.20 is 72 % of 485. Therefore, the discount is: 100% – 72% = 28%.

15) Answer: B

$average = \frac{sum\ of\ terms}{number\ of\ terms} \Rightarrow average = \frac{(38+24+32+20)}{4} \Rightarrow \frac{114}{4} = 28.5$

16) Answer: B

Let x be the number. Write the equation and solve for x.

$\frac{(56-x)}{x} = 7$ (cross multiply)

$(56 - x) = 7x$, then add x both sides. $56 = 8x$, now divide both sides by 8. $\Rightarrow x = 7$.

17) Answer: C

Use Pythagorean Theorem: $a^2 + b^2 = c^2$

$10^2 + 24^2 = C^2 \Rightarrow 100 + 576 = C^2 \Rightarrow 676 = c^2 \Rightarrow c = 26$

18) Answer: A

$19 \times 10 = \$190$

Petrol use: $7 \div 2 = 3.5, 10 \times 3.5 = 35$ liters

Petrol cost: $35 \times \$2.5 = \$87.5$

Money earned: $\$190 - \$87.5 = \$102.5$

19) Answer: D

If the length of the box is 100, then the width of the box is one fifth of it, 20, and the height of the box is 2 (one tenth of the width). The volume of the box is:

$V = lwh = (100)(20)(2) = 4,000$

20) Answer: B

Let x be the original price.

If the price of the sofa is decreased by 30% to \$315, then: $70\%\ of\ x = 315 \Rightarrow 0.7x = 315 \Rightarrow x = 315 \div 0.7 = 450$

TASC Subject Test – Mathematics

21) Answer: C

The percent of girls playing tennis is: $52\% \times 35\% = 0.52 \times 0.35 = 0.182 = 18.2\%$

22) Answer: C

Use this formula: Percent of Change

$$\frac{New\ Value - Old\ Value}{Old\ Value} \times 100\ \%$$

$\frac{72,000-90,000}{90,000} \times 100\ \% = -20\ \%$ and $\frac{57,600-72,000}{72,000} \times 100\% = -20\ \%$

23) Answer: D

Let x be the smallest number. Then, these are the numbers:

$x, x+1, x+2, x+3, x+4$

Average $= \frac{sum\ of\ terms}{number\ of\ terms} \Rightarrow 112 = \frac{x+(x+1)+(x+2)+(x+3)+(x+4)}{5} \Rightarrow 112 = \frac{5x+10}{5}$

$\Rightarrow 112 = x + 2 \Rightarrow x = 110$

24) Answer: C

Area of the circle is less than 144π. Use the formula of areas of circles.

$$Area = \pi r^2 \Rightarrow \pi r^2 < 144\pi \Rightarrow r^2 < 144 \Rightarrow r < 12$$

Radius of the circle is less than 12. Let's put 6 for the radius. Now, use the circumference formula:

$$Circumference = 2\pi r = 2\pi (12) = 24\pi$$

Since the radius of the circle is less than 12. Then, the circumference of the circle must be less than 24π. Choices C is less than 24π

25) Answer: C

Use simple interest formula: $I = prt$ (I= interest, p = principal, r = rate, t = time)

$$I = (10,000)(0.0225)(4) = 900$$

26) Answer: A

Add the first 3 numbers. $70 + 77 + 80 = 227$

To find the distance traveled in the next 5 hours, multiply the average by number of hours. Distance $= Average \times Rate = 65 \times 5 = 325$

Add both numbers. $227 + 325 = 552$

WWW.MathNotion.Com

TASC Subject Test – Mathematics

27) Answer: C

The ratio of boy to girls is 9: 5. Therefore, there are 9 boys out of 14 students. To find the answer, first divide the total number of students by 14, then multiply the result by 9.

$490 \div 14 = 35 \Rightarrow 35 \times 9 = 315$

28) Answer: D

The area of the trapezoid is: $Area = \frac{1}{2}h(b_1 + b_2) = \frac{1}{2}(x)(15 + 20) = 210$

$\rightarrow 35x = 420 \rightarrow x = 12$

$y = \sqrt{5^2 + 12^2} = \sqrt{25 + 144} = \sqrt{169} = 13$

The perimeter of the trapezoid is: $12 + 15 + 13 + 20 = 60$

29) Answer: C

The equation of a line is in the form of $y = mx + b$, where m is the slope of the line and b is the $y - intercept$ of the line.

Two points $(-1, -4)$ and $(2, 5)$ are on the line A. Therefore, the slope of the line A is:

$slope\ of\ line\ A = \frac{y_2 - y_1}{x_2 - x_1} = \frac{5-(-4)}{2-(-1)} = \frac{9}{3} = 3$

The slope of line A is 3. Thus, the formula of the line A is:

$y = mx + b = 3x + b$, choose a point and plug in the values of x and y in the equation to solve for b. Let's choose point $(2, 5)$. Then:

$$y = 3x + b \rightarrow 5 = 6 + b \rightarrow b = 5 - 6 = -1$$

The equation of line A is: $y = 3x - 1$

Now, let's review the choices provided:

A. $(2, 9)$ $y = 3x - 1 \rightarrow 9 = 6 - 1 = 5,$ This is not true.

B. $(-2, 5)$ $y = 3x - 1 \rightarrow 5 = -6 - 1 = -7,$ This is not true.

C. $(4, 11)$ $y = 3x - 1 \rightarrow 11 = 12 - 1 = 11,$ This is true!

D. $(0, 2)$ $y = 3x - 1 \rightarrow 2 = 0 - 1 = -1,$ This is not true.

30) Answer: D

12% of the volume of the solution is alcohol. Let x be the volume of the solution. Then:

12% of $x = 21.6$ ml

TASC Subject Test – Mathematics

$0.12\ x = 21.6 \Rightarrow \frac{12x}{100} = \frac{216}{10}$ cross multiply

$120x = 21,600 \Rightarrow$ (devide by 120) $x = 180$

31) Answer: B

$average = \frac{sum\ of\ terms}{number\ of\ terms}$

The sum of the high of all constructions is: $15 \times 140 = 2,100$ m

The sum of the high of all towers is: $5 \times 120 = 600$ m

The sum of the high of all building is: $2,100 + 600 = 2,700$

$average = \frac{2,700}{20} = 135$

32) Answer: C

Let x be the original price.

If the price of a laptop is decreased by 15% to $170, then:

$85\ \%\ of\ x = 170 \Rightarrow 0.85x = 170 \Rightarrow x = 170 \div 0.85 = 200$

33) Answer: A

Let x be the number of years. Therefore, $1,500 per year equals $1,500x$.

starting from $22,700 annual salary means you should add that amount to $1,500x$.

Income more than that is: I > $1,500x + 22,700$

34) Answer: D

Use the information provided in the question to draw the shape.

Use Pythagorean Theorem: $a^2 + b^2 = c^2 \Rightarrow 20^2 + 15^2 = c^2$

$\Rightarrow 400 + 225 = c^2 \Rightarrow 625 = c^2 \Rightarrow c = 25$.

35) Answer: D

Write the equation and solve for M:

$0.64\ F = 0.32\ M$, divide both sides by 0.32, then: $\frac{0.64}{0.32} F = M$, therefore:

M = 2 F, and M is 2 times of F or it's 200% of F.

36) Answer: B

To find the number of possible outfit combinations, multiply number of options for each factor: $9 \times 5 \times 7 = 315$

TASC Subject Test – Mathematics

37) Answer: C

Formula for the Surface area of a cylinder is:

$SA = 2\pi r^2 + 2\pi rh \rightarrow 72\pi = 2\pi r^2 + 2\pi r(5) \rightarrow r^2 + 5r - 36 = 0$

$(r + 9)(r - 4) = 0 \rightarrow r = 4 \quad \text{or} \quad r = -9 \ (unacceptable)$

38) Answer: B

Use distance formula:

$Distance = Rate \times time \Rightarrow 338 = 52 \times T$, divide both sides by 52. $\Rightarrow T = 6.5$ hours.

Change hours to minutes for the decimal part. $0.5 \ hours = 0.5 \times 60 = 30 \ minutes$.

39) Answer: B

To find the discount, multiply the number by (100% – rate of discount).

Therefore, for the first discount we get: (500) (100% – 30%) = (500) (0.70)

For the next 10% discount: (500) (0.70) (0.90).

40) Answer: A

Let's compare each fraction:

$\frac{1}{8} < \frac{3}{14} < \frac{2}{9}$

Only choice A provides the right order.

TASC Subject Test – Mathematics

TASC Practice Test 1
Section 2 - No Calculator

41) Answer: 140

The question is this: 3.15 is what percent of 2.25?

Use percent formula:

$$part = \frac{percent}{100} \times whole$$

$3.15 = \frac{percent}{100} \times 2.25 \Rightarrow 3.15 = \frac{percent \times 2.25}{100} \Rightarrow 315 = percent \times 2.25 \Rightarrow percent = \frac{315}{2.25} = 140.$

42) Answer: 11.25

$5x - 9 = 6 \to 5x = 6 + 9 = 15 \to x = \frac{15}{5} = 3$

Then; $4x - \frac{3}{4} = 4(3) - \frac{3}{4} = 12 - 0.75 = 11.25.$

43) Answer: 98

First draw an isosceles triangle. Remember that two sides of the triangle are equal. Let put a for the legs. Then:

$a = 14 \Rightarrow$ area of the triangle is $= \frac{1}{2}(14 \times 14) = \frac{196}{2} = 98 \ cm^2.$

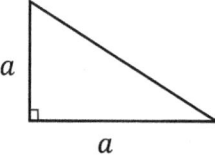

Isosceles right triangle

44) Answer: 18

The input value is 3. Then: $x = 3$

$f(x) = 3x^2 - 7x + 12 \to f(3) = 3(3^2) - 7(3) + 12 = 27 - 21 + 12 = 18$

45) Answer: 0.15

Write the ratio of $4a$ to $9b$.

$\frac{4a}{9b} = \frac{1}{15}$

Use reciprocal of $\frac{4}{9}$ and multiply both sides, then simplify.

$\frac{4a}{9b} = \frac{1}{15} \Rightarrow \frac{9}{4} \times \frac{4a}{9b} = \frac{1}{15} \times \frac{9}{4} \Rightarrow \frac{a}{b} = \frac{9}{60} = 0.15$

TASC Subject Test – Mathematics

46) Answer: 9 m

The rate of construction company = $\frac{14 \text{ cm}}{1 \text{ min}}$ = 14 cm/min

Height of the wall after 50 minutes = $\frac{14 \text{ cm}}{1 \text{ min}} \times 50$ min = 700 cm

Let x be the height of wall, then $\frac{7}{9}x = 700$ cm → $x = \frac{9 \times 700}{7}$ → $x = 900$ cm = $9\,m$

47) Answer: 10

Use PEMDAS (order of operation):

$8 \times (-5) + 10 - 5(-4 - 14 \times 6) \div 11 = -40 + 10 - 5(-4 - 84) \div 11 = -30 - 5(-88) \div 11 = -30 + 440 \div 11 = -30 + 40 = 10$.

48) Answer: 160.

The perimeter of the trapezoid is 52.

Therefore, the missing side (height) is = $52 - (16 + 9 + 11) = 16$

Area of a trapezoid: $A = \frac{1}{2} h (b1 + b2) = \frac{1}{2} (16)(9 + 11) = 160$

49) Answer: $2M - 12$

$x + y = M$ and $x = 6 \Rightarrow 6 + y = M \Rightarrow y = M - 6, \Rightarrow 2y = 2(M - 6) = 2M - 12$

50) Answer: 2

Let x be the length of an edge of cube, then the volume of a cube is: $V = x^3$

The surface area of cube is: $SA = 6x^2$

The volume of cube A is $\frac{1}{3}$ of its surface area. Then:

$x^3 = \frac{6x^2}{3} \to x^3 = 2x^2$, divide both side of the equation by x^2. Then:

$$\frac{x^3}{x^2} = \frac{2x^2}{x^2} \to x = 2$$

51) Answer: -7

Since $N = -6$, substitute -6 for N in the equation $\frac{3x-3}{4} = N$, which gives $\frac{3x-3}{4} = -6$.

Multiplying both sides by 4 gives $3x - 3 = -24$ and then adding 3 to both sides, then $3x = -21$ then, Dividing both sides by 3 then, $x = -7$.

52) Answer: $15\sqrt{2}$

The relationship among all sides of special right triangle 45°; 45°; 90° is provided in this triangle:

In this triangle, the opposite side of 45° angle is half of the hypotenuse.

Draw the shape of this question.

The latter is the hypotenuse. Therefore, the latter is $15\sqrt{2}$ feet.

TASC Subject Test – Mathematics

Practice Test 2

Section 1 - Calculator

1) Answer: B

The area of the square is 547.56. Therefore, the side of the square is square root of the area. $\sqrt{547.56} = 23.4$

Four times the side of the square is the perimeter: $4 \times 23.4 = 93.6$

2) Answer: D

Write the equation and solve for B: $0.57\,A = 0.19\,B$, divide both sides by 0.19, then: $\frac{0.57}{0.19} A = B$, therefore: $B = 3\,A$, and B is 3 times of A or it's 300% of A.

3) Answer: C

Simplify and combine like terms.

$(8x^4 - 4x^2 - 3x^5) - (3x^2 - 2x^5 + 5x^4)$

$\Rightarrow (8x^4 - 4x^2 - 3x^5) - 3x^2 + 2x^5 - 5x^4$

$\Rightarrow -x^5 + 3x^4 - 7x^2 = -(x^5 - 3x^4 + 7x^2)$.

4) Answer: A

2,400 out of 55,200 equals to $\frac{2,400}{55,200} = \frac{24}{552} = \frac{8}{184} = \frac{1}{23}$

5) Answer: D

Write the numbers in order: 4, 9, 17, 25, 28, 34, 51

Median is the number in the middle. So, the median is 25.

6) Answer: B

Use simple interest formula: $I = prt$ (I = interest, p = principal, r = rate, t = time)

$I = (16,000)(0.0165)(4) = 1,056$

7) Answer: D

Three times of 22,000 is 66,000. One sixth of them cancelled their tickets.

One sixth of 66,000 equal 11,000 ($\frac{1}{6} \times 66,000 = 11,000$).

$(66,000 - 11,000 = 55,000)$ fans are attending this week

WWW.MathNotion.Com

TASC Subject Test – Mathematics

8) Answer: B

the population is increased by 6% and 15%. 6% increase changes the population to 106% of original population. For the second increase, multiply the result by 115%.

$(1.06) \times (1.15) = 1.219 = 121.9\%$

21.9 percent of the population is increased after two years.

9) Answer: C

average (mean) $= \frac{\text{sum of terms}}{\text{number of terms}} \Rightarrow 35 = \frac{\text{sum of terms}}{30} \Rightarrow$ sum $= 35 \times 30 = 1,050$

The difference of 35 and 20 is 15. Therefore, 15 should be subtracted from the sum.

$1,050 - 15 = 1,035$

mean $= \frac{\text{sum of terms}}{\text{number of terms}} \Rightarrow$ mean $= \frac{1,035}{30} = 34.5$

10) Answer: A

Solve for x.

$-2 \leq 5x - 7 < 3 \Rightarrow$ (add 7 all sides) $-2 + 7 \leq 5x - 7 + 7 < 3 + 7 \Rightarrow 5 \leq 5x < 10$

$\Rightarrow$ (divide all sides by 5) $5 \leq 5x < 10$

x is between 1 and 2. Choice A represent this inequality.

11) Answer: D

To get a sum of 6 for two dice, we can get 5 different options:

$(1, 5), (5, 1), (2, 4), (4, 2), (3, 3)$

To get a sum of 9 for two dice, we can get 4 different options:

$(3, 6), (6, 3), (4, 5), (5, 4)$

Therefore, there are 9 options to get the sum of 6 or 9.

Since, we have $6 \times 6 = 36$ total options, the probability of getting a sum of 6 or 9 is 9 out of 36 or $\frac{9}{36} = \frac{1}{4}$.

12) Answer: C

Change the numbers to decimal and then compare.

$\frac{1}{8} = 0.125$; 0.4; $9\% = 0.09$; $\frac{1}{12} = 0.083$

Therefore $\frac{1}{12} < 9\% < \frac{1}{8} < 0.4$.

TASC Subject Test – Mathematics

13) Answer: B

The diagonal of the square is 8. Let x be the side.

Use Pythagorean Theorem: $a^2 + b^2 = c^2$

$x^2 + x^2 = 8^2 \Rightarrow 2x^2 = 64 \Rightarrow x^2 = 32 \Rightarrow x = \sqrt{32}$

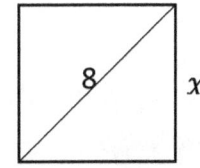

The area of the square is: $\sqrt{32} \times \sqrt{32} = 32$

14) Answer: A

Volume of a box = length × width × height = $3 \times 5 \times 11 = 165$

15) Answer: C

average = $\frac{\text{sum of terms}}{\text{number of terms}} \Rightarrow 29 = \frac{\text{sum of 5 numbers}}{5} \Rightarrow$ sum of 5 numbers = $29 \times 5 = 145$

$45 = \frac{\text{sum of 3 numbers}}{3} \Rightarrow$ sum of 3 numbers = $3 \times 45 = 135$

sum of 5 numbers − sum of 3 numbers = sum of 2 numbers

$145 - 135 = 10 \rightarrow$ average of 2 numbers = $\frac{10}{2} = 5$.

16) Answer: D

Solving Systems of Equations by Elimination

$\begin{array}{l} 4x + 5y = -11 \\ 3x - 2y = 9 \end{array}$ Multiply the first equation by 3, and second equation by −4, then add two equations.

$\begin{array}{l} 3(4x + 5y = -11) \\ -4(3x - 2y = 9) \end{array} \Rightarrow \begin{array}{l} 12x + 15y = -33 \\ -12x + 8y = -36 \end{array} \Rightarrow 23y = -69 \Rightarrow y = -3$.

17) Answer: A

The width of the rectangle is triple its length. Let x be the length. Then, $width = 3x$

Perimeter of the rectangle is 2 (width + length) = $2(3x + x) = 32 \Rightarrow 8x = 32 \Rightarrow x = 4$

Length of the rectangle is 4 meters.

18) Answer: B

In the stadium the ratio of home fans to visiting fans in a crowd is 4:7. Therefore, total number of fans must be divisible by 11: $4 + 7 = 11$.

Let's review the choices:

A. 38,980: $38,980 \div 11 = 3,543.63$

TASC Subject Test – Mathematics

B. 47,300:   47,300 ÷ 11 = 4,300

C. 41,500:   41,500 ÷ 11 = 3,772.72

D. 53,670:   53,670 ÷ 11 = 4,879.09

Only choice B when divided by 11 result a whole number.

19) Answer: C

Plug in 86 for F and then solve for C.

$C = \frac{5}{9}(F - 32) \Rightarrow C = \frac{5}{9}(86 - 32) \Rightarrow C = \frac{5}{9}(54) = 30.$

20) Answer: C

$3x - 2y = -8$. Plug in the values of x and y from choices provided. Then:

A. $(-1, 3)$: $3(-1) - 2(3) = -3 - 6 = -9$,   This is NOT true!

B. $(2, -2)$: $3(2) - 2(-2) = 6 + 4 = 10$,   This is NOT true!

C. $(0, 4)$: $3(0) - 2(4) = 0 - 8 = -8$,   This is true!

D. $(5, 1)$: $3(5) - 2(1) = 15 - 2 = 13$,   This is NOT true!

21) Answer: D

To find the number of possible outfit combinations, multiply number of options for each factor: $4 \times 10 \times 6 = 240$

22) Answer: D

$\text{Probability} = \frac{\text{number of desired outcomes}}{\text{number of total outcomes}} = \frac{15}{32+25+15+28} = \frac{15}{100} = \frac{3}{20}$

23) Answer: A

First, find the sum of four numbers.

$\text{Average} = \frac{\text{sum of terms}}{\text{number of terms}} \Rightarrow 42 = \frac{\text{sum of 4 numbers}}{5} \Rightarrow \text{sum of 4 numbers} = 5 \times 42 = 210$

The sum of 5 numbers is 210. If a sixth number that is greater than 60 is added to these numbers, then the sum of 6 numbers must be greater than 270 ($210 + 60 = 270$).

If the number was 60, then the average of the numbers is:

$\text{average} = \frac{\text{sum of terms}}{\text{number of terms}} = \frac{270}{6} = 45$

Since the number is bigger than 60. Then, the average of six numbers must be greater than 45. Choices A is greater than 46.

TASC Subject Test – Mathematics

24) Answer: A

The ratio of boy to girls is 2:5. Therefore, there are 2 boys out of 7 students. To find the answer, first divide the total number of students by 7, then multiply the result by 2.

$63 \div 7 = 9 \Rightarrow 9 \times 2 = 18$

There are 18 boys and 45 (63 − 18) girls. So, 27 more boys should be enrolled to make the ratio 1:1

25) Answer: B

The perimeter of the trapezoid is 63 cm.

Therefore, the missing side (height) is = 63 − (20 + 15 + 12) = 16

Area of a trapezoid: $A = \frac{1}{2} h (b_1 + b_2) = \frac{1}{2} (16) (12 + 15) = 216$

26) Answer: B

The probability of choosing spades is $\frac{13}{52} = \frac{1}{4}$

27) Answer: D

Surface Area of a cylinder = $2\pi r (r + h)$,

The radius of the cylinder is 5 (10 ÷ 2) inches, and its height is 20 inches. Therefore,

Surface Area of a cylinder = $2\pi (5) (5 + 20) = 250\pi$

28) Answer: C

Simplify.

$2x^3 y^6 (-5xy^4)^2 = 2x^3 y^6 (25x^2 y^8) = 50 x^5 y^{14}$

29) Answer: C

The distance between Daniel and Noa is 24 miles. Daniel running at 5.2 miles per hour and Noa is running at the speed of 4 miles per hour. Therefore, every hour the distance is 1.2 miles less. $24 \div 1.2 = 20$.

30) Answer: B

Let x be the number of new shoes the team can purchase. Therefore, the team can purchase 98 x.

TASC Subject Test – Mathematics

The team had $35,000 and spent $18,000. Now the team can spend on new shoes $17,000 at most.

Now, write the inequality: $98x + 18,000 \leq 35,000$

31) Answer: D

Let x be the number. Write the equation and solve for x.

70% of $x = 35 \Rightarrow 0.70\,x = 35 \Rightarrow x = 35 \div 0.70 = 50$

32) Answer: B

$\frac{48}{243}$, simplify by 3, then the number is the square root of $\frac{16}{81}$: $\sqrt{\frac{16}{81}} = \frac{4}{9}$

The cube of the number is: $(\frac{4}{9})^3 = \frac{64}{729}$

33) Answer: C

The failing rate is 24 out of 80, $\frac{24}{80}$

Change the fraction to percent:

$\frac{24}{80} \times 100\% = 30\%$

30 percent of students failed. Therefore, 70 percent of students passed the exam.

34) Answer: A

Let x be all expenses, then $\frac{30}{100}x = \$1,200 \rightarrow x = \frac{100 \times \$1,200}{30} = \$4,000$

He spent for his rent: $\frac{25}{100} \times \$4,000 = \$1,000$

35) Answer: C

Isolate and solve for x.

$\frac{2}{5}x + \frac{1}{24} = \frac{1}{8} \Rightarrow \frac{2}{5}x = \frac{1}{8} - \frac{1}{24} = \frac{1}{12} \Rightarrow \frac{2}{5}x = \frac{1}{12}$

Multiply both sides by the reciprocal of the coefficient of x.

$(\frac{5}{2})\frac{2}{5}x = \frac{1}{12}(\frac{5}{2}) \Rightarrow x = \frac{5}{24}$

36) Answer: C

First, find the number.

Let x be the number. Write the equation and solve for x.

TASC Subject Test – Mathematics

140% of a number is 210, then:

$1.4 \times x = 210 \Rightarrow x = 210 \div 1.4 = 150$

40% of 150 is: $0.4 \times 150 = 60$

37) Answer: A

Mrs. Thomson needs an 82% average to pass for five exams. Therefore, the sum of 5 exams must be at lease $5 \times 82 = 410$

The sum of 4 exams is: $72 + 76 + 85 + 93 = 326$

The minimum score Mrs. Thomson can earn on her fifth and final test to pass is: $410 - 326 = 84$

38) Answer: C

Let L be the length of the rectangular and W be the width of the rectangular.

Then, $L = 3W - 2$

The perimeter of the rectangle is 52 meters. Therefore: $2L + 2W = 52$

$$L + W = 26$$

Replace the value of L from the first equation into the second equation and solve for W:

$$(3W - 2) + W = 26 \rightarrow 4W - 2 = 26 \rightarrow 4W = 28 \rightarrow W = 7$$

The width of the rectangle is 7 meters, and its length is:

$$L = 4W - 2 = 4(7) - 2 = 26$$

The area of the rectangle is: length × width = 26 × 7 = 182

39) Answer: B

Use formula of rectangle prism volume.

V = (length) (width) (height) $\Rightarrow$ 4,320 = (24) (15) (height) $\Rightarrow$ height = 4,320 ÷ 360 = 12

40) Answer: D

Use Pythagorean Theorem: $a^2 + b^2 = c^2$

$12^2 + 16^2 = c^2 \Rightarrow 400 = c^2 \Rightarrow c = 20$

TASC Subject Test – Mathematics

TASC Practice Test 2

Section 2- No Calculator

41) Answer: 2.5

Write a proportion and solve for the missing number.

$\frac{70}{25} = \frac{7}{x} \rightarrow 70x = 7 \times 25 = 175$

$70x = 175 \rightarrow x = \frac{175}{70} = 2.5$.

42) Answer: 21

Plug in the value of x and y. $5(2x + 3y) + (14 - 5x)^2$ when $x = 2$ and $y = -1$

$= 5(2x + 3y) + (14 - 5x)^2 = 5(2(2) + 3(-1)) + (14 - 5(2))^2 = 5(1) + (4)^2$

$= 5 + 16 = 21$

43) Answer: −8

The equation of a line in slope intercept form is: $y = mx + b$; Solve for y.

$x - 8y = 9 \rightarrow -8y = -x + 9$; Divide both sides by (-8). Then: $y = \frac{1}{8}x - \frac{9}{8}$

The slope of this line is $\frac{1}{8}$.

The product of the slopes of two perpendicular lines is -1. Therefore, the slope of a line that is perpendicular to this line is:

$m_1 \times m_2 = -1 \Rightarrow \frac{1}{8} \times m_2 = -1 \Rightarrow m_2 = \frac{-1}{\frac{1}{8}} = -8$

44) Answer: –71

Use PEMDAS (order of operation):

$[4 \times (-16)] + ([(-5) \times (-16)] \div 5) + (-23) = [-64] + (80 \div 5) - 23 = -64 + 16 - 23 = -71$.

45) Answer: −56

To solve absolute values equations, write two equations.

$-4x + 2$ can equal positive 30, or negative 30. Therefore,

$-4x + 2 = 30 \Rightarrow -4x = 30 - 2 = 28 \Rightarrow x = -7$

$-4x + 2 = -30 \Rightarrow -4x = -30 - 2 = -32 \Rightarrow x = 8$

Find the product of solutions: $-7 \times 8 = -56$

TASC Subject Test – Mathematics

46) Answer: 116

$-35 = 81 - x$; First, subtract 81 from both sides of the equation. Then:

$-35 - 81 = 81 - 81 - x \rightarrow -116 = -x$

Multiply both sides by (-1): $\rightarrow x = 116$

47) Answer: 8.5

$7x - 25 = -4 \rightarrow 7x = 25 - 4 = 21 \rightarrow x = \frac{21}{7} = 3$

Then, $3x - \frac{1}{2} = 3(3) - 0.5 = 9 - 0.5 = 8.5$

48) Answer: 1,728

If the length of the box is 48, then the width of the box is one fourth of it, 12, and the height of the box is 3 (one fourth of the width). The volume of the box is:

$V = lwh = (48)(12)(3) = 1,728$

49) Answer: 14

Let y be the width of the rectangle. Then; $16 \times y = 224 \rightarrow y = \frac{224}{16} = 14$

50) Answer: 75%

Number of males in classroom is: $60 - 15 = 45$

Then, the percentage of males in the classroom is: $\frac{45}{60} \times 100 = 0.75 \times 100 = 75\%$

51) Answer: 35

Let x be the number. Write the equation and solve for x.

$\frac{1}{9} \times 135 = \frac{3}{7}x \rightarrow \frac{1 \times 135}{9} = \frac{3x}{7}$, use cross multiplication to solve for x.

$7 \times 135 = 3x \times 9 \Rightarrow 945 = 27x \Rightarrow x = 35$

52) Answer: 44.44

Average $= \frac{\text{sum of terms}}{\text{number of terms}}$

The sum of the weight of all girls is: $27 \times 38 = 1,026$ kg

The sum of the weight of all boys is: $23 \times 52 = 1,196$ kg

The sum of the weight of all students is: $1,026 + 1,196 = 2,222$ kg

Average $= \frac{2,222}{50} = 44.44$

"End"

www.ingramcontent.com/pod-product-compliance
Lightning Source LLC
Chambersburg PA
CBHW080438110426
42743CB00016B/3200